W0042596

Visualization in
Scientific Computing '95

Proceedings of the Eurographics Workshop
in Chia, Italy, May 3–5, 1995

R. Scateni
J. van Wijk
P. Zanarini (eds.)

Eurographics

SpringerWienNewYork

Dr. Riccardo Scateni
CRS4, Cagliari, Italy

Dr. Jarke J. van Wijk
ECN, Petten, The Netherlands

Dr. Pietro Zanarini
CRS4, Cagliari, Italy

This work is subject to copyright.
All rights are reserved, whether the whole or part of the material is concerned, specifically those of translation, reprinting, re-use of illustrations, broadcasting, reproduction by photocopying machines or similar means, and storage in data banks.

© 1995 Springer-Verlag/Wien

Typesetting: Camera ready by editors and authors

Graphic design: Ecke Bonk

Printed on acid-free and chlorine-free bleached paper

With 110 partly coloured Figures

ISSN 0946-2767
ISBN-13:978-3-211-82729-1   e-ISBN-13:978-3-7091-9425-6
DOI: 10.1007/978-3-7091-9425-6

Preface

Visualization is nowadays indispensable to get insight into the huge amounts of data produced by large scale simulations or advanced measurement devices. The use of computer graphics for scientific purposes has become a well established discipline, known as Scientific Visualization. Many problems still have to be solved, and hence the field is a very active area for research and development. This book represents results of the sixth in a well established series of international workshops on *Visualization in Scientific Computing* organized by the EUROGRAPHICS Association in collaboration with CRS4 (Center for Advanced Studies, Research and Development in Sardinia), held from May 3 to May 5, 1995, in Chia, Italy.

The thirteen contributions selected for this volume cover a wide range of topics, ranging from detailed algorithmic studies to searches for new metaphors. A rough division can be made into the parts *interaction*, *irregular meshes*, *volume rendering*, and *applications*.

Interaction in three dimensions is a challenging area for research. The use of three-dimensional user interfaces for more natural manipulation of three-dimensional data and their visualization is natural, but is far from trivial to realize. Pang et al. investigate the use of common objects such as spray cans and carving knives as metaphors for visualization tools, in order to provide an intuitive and natural three dimensional user interface. Gibson uses a voxel-based data representation, not only for visualization, but also for physical modeling of objects. A prototype system under development for haptic exploration is discussed.

The first systems for visualization assumed that researchers were making use of cartesian, regular grids. This is a gross simplification. Tetrahedral, adaptive, hierarchical, and many other types of *irregular meshes* are used for advanced numerical simulations. This has been recognized in current visualization research, and is studied intensively. Schmidt et al. present a visualization system that works with any type of unstructured grid. Rumpf et al. propose a unified approach to visualize data from arbitrary meshes, using a procedural interface to the mesh.

The aim of *volume rendering* is to visualize three-dimensional data-sets without intermediate representations. It was used originally for medical applications, typically the visualization of density volumes obtained from stacking computer tomography slices. Although much progress has been made, the results can still be improved. The use of irregular meshes poses additional problems. Fruehauf shows how to visualize opaque isosurfaces in irregular grids using ray-casting. Cignoni et al. consider projective volume rendering of tetrahedral tesselations. Several algorithms are compared, and a new method for tetrahedral projection is presented. Rau et al. describe how irregular samples, not located on a structured mesh, can be visualized via a forward mapping technique. The kernel of the reconstruction filter to be used is discussed in detail. Grosso et al. also discuss reconstruction filters, but now for frequency domain volume rendering. Wavelets are used for reconstruction as well as for hierarchical visualization, in order to gain performance. Zhang et al. use a system for parallel volume rendering on networked workstations for improving the performance. The order of pixel traversal is one of the critical steps. Pixel traversal via a space filling curve gives the best results.

Scientific Visualization is an application-driven discipline. Hence the last four papers cover interesting *applications* of visualization to the solution of more particular problems. The visualization of local stability of dynamical systems is discussed by Fischel et al. Many different techniques are used to get insight into the complex behavior resulting from seemingly simple differential equations. Mulder et al. show how the third dimension can be used to visualize the time evolution of two-dimensional results. This logging facility gives much additional insight, especially in a computational steering context. Leone et al. use visualization for internal combustion simulations. They show how several scalar and vector fields can be visualized simultaneously. De Leeuw et al. use spot noise to generate images of the results of numerical flow simulation that can be compared directly with images taken during wind-tunnel experiments. Such comparative visualization of data from different sources provides more insight in similarities and differences.

The editors want to thank everybody involved in the production of this book. In particular we thank all contributing authors, the members of the Program Committee, and the staff of Springer-Verlag, Wien, for the smooth cooperation.

Finally, we hope to have provided the readers with a selection of state-of-the-art results and techniques in Scientific Visualization research, which they can use to find solutions for their visualization problems.

Riccardo Scateni
Jarke J. van Wijk
Pietro Zanarini

Contents

Metaphors for visualization

Alex Pang and Michael Clifton

Computer and Information Sciences Board
University of California, Santa Cruz
California, 95064, USA

Abstract

This paper investigates the use of several common objects such as spray cans, flashlights, carving knives, and others as metaphors for visualization. The motivation behind this work is to provide an intuitive and natural 3D interface such that users will view these objects as tools during the visualization process. To help achieve this goal, the selection of the object tools are derived from expressions commonly used during visualization. For example, "let's paint this (iso)surface red"; "cut away the front part of that volume"; "look at it from this angle"; "you get similar effects from an x-ray photo"; and "let's explore this data set". In this paper, we also extend the metaphor of cutting planes to allow users to carve non-planar cross-sectional cuts through their data sets. This extension will directly benefit applications such as medical visualization where one might want to generate curved coronal cross-sections of lumbar spines; and in oceanography where one might want to compare numerical model output against data obtained along non-planar ship tracks.

1 Introduction

The goal of visualization is to present data in such a way as to facilitate its understanding. The amount and ease in which insight is gained from the data set is an indication of the success of the visualization process. Equally as important as the development of powerful visualization methods is the development of techniques and interface methods that make these powerful tools available to the users. The tools should be made accessible in an unencumbered, direct, and natural way – first to gain acceptance from the users, and second to promote unimpeded investigation of their data without having to think about the visualization methods. This is not to say that the users do not need to be aware of the limitations of the methods, but rather the interface should be so natural to use that it is almost invisible to the users. In this case, the measure of an interface's success is its apparent absence from the user's thought process while investigating their data set.

Our approach at achieving such an interface is to take advantage of common adages or metaphors used in discussions when examining data sets. These metaphors are then translated to common day tools and devices for the users as interaction handles. Thus, users are provided with visualization tools to directly manipulate when examining their data. These tools are identified from colloquialisms such as: "let's paint this (iso)surface red"; "cut away the front part of that volume"; "look at it from this angle"; "you get similar effects from an x-ray photo"; and "let's explore this data set". Clearly, from the visualization users' point of view, they are not thinking in terms of the feature extraction or rendering algorithm but rather the physical manipulation needed to get the desired effect. We can enhance this train of thought by providing users with virtual 3D widgets that they can directly manipulate and which closely correspond to their mental model of manipulating their data sets. In the above examples, the virtual 3D widgets might respectively correspond to a spray can or the low tech paint brush; butcher knife or carving knife; a camera; a portable x-ray camera or perhaps a space age laser gun of some sort; exploration may conjure groping around in the dark. So, we can provide a virtual flashlight.

Since these metaphors are, in almost all cases, 3D objects, it is natural to implement these with virtual reality input devices (gloves and trackers, at a minimum). The 3D widgets that correspond to these objects differ from those employed in most applications of virtual reality technology to scientific visualization in their means of interaction. The goal is to design the interaction with the 3D widgets in such a way that the user will not have to consciously think about how to interact with the widget. This frees up the mental process of the user to concentrate on data analyses instead of keeping track of how to interact with the widget in order to get to the data. This objective is also in line with the human-computer software interface recommendations by [1] and the scenarios of use approach by [2]. The added level of indirection that users have to deal with is surprisingly common in application of virtual reality technology to scientific visualization applications. For example, one of the 3D widgets [3, 4] presented was a floating rectangle where the corners, edges and the entire rectangle were grabbable. The position, size and orientation of this rectangle was then controlled by grabbing on to the different parts of the rectangle and manipulating them. While the different components of the rectangle are highlighted to alert the user that they are grabbable, the user still has to think in terms of grabbing corners or edges of the rectangle and manipulating those to get to the data. Thus, there is an added level of interaction for the user to deal with when trying to visualize their data sets. In contrast, the 3D widgets in this paper rely on the users' experience with common everyday objects. Thus, control and manipulation of these 3D widgets can be achieved in a natural and unobtrusive way. A simple example, based on [5], is the use of a treadmill/bicycle metaphor to allow the user to walk/bike around the data set. In this scenario, the bike handles are used to control the direction of travel and view while the leg movements control the speed. This type of metaphor allows the user to feel very comfortable and concentrate on looking at the data as opposed to consciously interacting with the widgets to get at the data.

2 Previous/Ongoing work

There are several notable application of virtual reality interfaces to aid scientific visualization. One that has received significant amounts of attention is the virtual wind tunnel experiment [6]. This is an immersive environment where the user can move about and examine flow fields by simply positioning a rake that generates streamlines. The naturalness of directly positioning a tool in the neighborhood of the data where visualization is needed, coupled with the interactive rates that this is achieved are the key ingredients of success in their work. The utility of these virtual reality interfaces can be greatly enhanced when combined with the power of using metaphors that facilitate their use. The recognition of gestures associated with some of these metaphors have also been investigated by [7].

The interactions associated with the metaphors that we are proposing will build upon and extend the conceptual design of spray rendering [8, 9]. Spray rendering uses the metaphor of providing the users with a shelf of spray cans, each of which can generate a different way of visualizing a data set. For example, a spray can might generate an iso-surface in a scalar field, while another spray can might generate streamlines in a flow field. An attractive feature of spray rendering is that most users have at one time or another operated an aerosol can. Hence, learning how to use the system is quite intuitive.

Internally, spray rendering combines particle systems [10] and behavioral animation [11]. Particle systems were originally designed for modeling objects that were difficult to model using traditional techniques but have also been used in visualization applications such as particle aging and tracing [12]. In these applications the particles were usually passive and were simply advected by the flow field. We incorporate behavioral animation strategies with these particles so that they may interact with other particles as well as with the data field that they are traveling through. These smart particles (sparts) have an endowed intelligence which can be broken down into roughly two parts: targets and behaviors. Targets are features, usually in the data set, that the spart is programmed to seek out. As these targets are found, the sparts are made visible according to how their behaviors are defined. For example, the marching cube [13] algorithm may be modified so that several independent sparts are looking for the iso-surface rather than have the marching cube algorithm process the entire volume data. In this case, the iso-surface sparts would be seeking for surfaces of the specified threshold value. Once these surfaces are found, the spart would leave a polygon which will then be sent to the renderer. It is also possible for the spart to leave behind a non-visible marker for communication with other cooperating sparts. One example where sparts work in pairs is the flow tracking ribbon sparts [14]. Using this framework, we have been able to modify several existing visualization techniques and also introduce some new ones [15], as well as extend it to a collaborative environment where multiple geographically distributed users can share spray cans and visualizations within a shared virtual workspace [16].

3   Approach

We are investigating the effectiveness of several metaphors for visualization that allow users to directly manipulate and generate visualization through interactions with VR devices. These metaphors include but are not limited to: spray cans, flashlights, and kitchen knives. Based on the chosen metaphor, the operations are then mapped to the VR devices so that the appropriate hand gestures and postures are interpreted properly. Outlined below are some of the preliminary postures and gestures associated with each metaphor and how they adjust the relevant parameters. The same postures and gestures would apply to the left hand as well as to the right hand. Note that while most VR implementations try to achieve immersion, this is not a requirement for the initial phase of this work. The same metaphors may later be applied to immersive or augmented environments.

3.1   Spray can

The figures below use some objects as props notably a $2 (dish washing) "cyber" glove to illustrate the different postures and gestures associated with the use of each metaphorical object.

It is important to note that this metaphor allows the user to pretend that s/he is holding the can (as in Fig. 2). This hand posture can then be mapped to control the different parameters of the virtual spray can.

Default posture: the can is held with the index finger on the nozzle ready to fire.

Pointing posture: the direction of spray is calculated based on the index finger orientation. It is roughly 45 degrees down from the first flexion joint of the index finger.

Firing gesture: the amount of index finger flexion is mapped to the number of sparts that are being sprayed. The sparts are sprayed continuously while the index finger is flexed.

Cone adjustment gesture: nozzle size can be increase/decreased by the amount of abduction between the thumb and the index finger.

Figure 1: Using the spray can.   Figure 2: Hand posture for holding can.

3.2   Flashlight

While the shape of the spray can has remained pretty universal, the flashlight has undergone more shape changes. Let's look at the traditional flashlight and the more modern design with a separate handle. Both require the user to hold the flashlight or the handle with 4 fingers clasp around it and the thumb poised over the on/off switch and pointing in the direction of the flashlight.

Pointing posture: controlled by thumb direction.

Firing gesture: on/off and the amount of sparts (or photons) to shoot is controlled by the thumb flexion.

Parameter adjustment postures: Since we are mapping non-flashlight parameters, such as beam shape and spread angle, the mapping here is not as natural. One possibility is to control beam shape and amount of spread angle by the amount of finger flexion.

Figure 3: Using the flashlight.   Figure 4: Posture for holding flash-light.

3.3   Knife

This metaphor allows the user to slice and dice their data sets in a natural and direct way. However, it also presents some interesting issues such as segmentation of hand gestures as pointed out in [17]. It also requires us to modify our local contouring algorithm to hand-carved surfaces. Since the hand posture for holding a knife is quite similar to holding a flashlight, we modified the hand posture to a karate chop instead. This way, the user can pretend that his/her hand is the knife itself with the cutting edge extending from base of the palm to the end of the small finger. Unwanted pieces of the sliced data may be successively removed by simply flicking the thumb. There are two ways of using the knife:

Traditional: This includes planar slicing, sweeping and rotation of the knife. In this case, the palm defines the cutting plane. The direction of the chop is defined by the forward movement with the cutting edge as the leading edge. A new cut is initiated when the user first moves the hand back and then moving forward again with the cutting edge. The cutting plane can be swept around the volume when the hand movement is normal to the cutting plane. Wrist movements control plane orientations.

Carving: Non-planar cross-sections can be defined in two ways:

1. Non-planar 3D ruled surfaces can be defined by simply moving the cutting edge along some curve while keeping the 4 fingers straight. This may be applied to taking curved coronal slices of cat-scan data or to curved slices around structures in CFD data.

2. Non-planar 3D curved surfaces can also be generated in a similar manner by allowing the user to flex the 4 fingers. The amount of finger flexion would define the amount of surface curvature.

Figure 5: Using the "Ginzu" knife. Figure 6: "The Chop". Hand posture to represent virtual knife.

3.4   Helmet/Eyeball-in-hand

The three metaphors listed above allow the user to manipulate the data set directly. We also need a mechanism for the user to navigate around the data set. There are several options here. Since our application is scientific visualization as opposed to creating virtual worlds, "flying" around the data set is not a necessity. What the user needs is a way to view the data from different positions and orientations; perhaps from within the data set as well. The metaphor we chose to adopt here is the eyeball-in-hand as proposed by [18]. This can be simply modified to an eyeball or flashlight mounted on a spelunker's helmet when we move to an immersive environment. In the meantime, the user uses one hand to manipulate the data and the other hand to control our virtual (tennis ball) eyeball.

Gaze posture: the hand position is mapped to the eye position; while the finger direction is mapped to the gaze direction. The middle finger provides the up vector.

New view: can be obtained by simply moving the hand and adjusting the wrist to the new gaze direction.

Note that aside from the technical aspects of gesture recognition, one must also be sensitive to the social implications of certain hand postures and gestures. For example, the okay (thumb forming a circle with the index finger) or the good luck (thumbs up) symbol may mean different things to different culture; much like the American way of waving goodbye may be construed to mean a beckoning

Figure 7: Holding the eyeball.   Figure 8: Hand posture for holding virtual eyeball.

gesture in other cultures [19]. Since one of the broader goals of spray rendering is collaborative visualization among scientists who are geographically distributed and potentially with different cultural background, it is important to take these subtleties into consideration.

4   Implementation

We have concentrated our initial efforts at implementing the knife as a metaphor for visualization. The instrumentats used in this work are: an 18 sensor Cyberglove, two 6D Ascension trackers, and Crystal Eyes stereo glasses.

Currently, we have implemented four methods for interacting with the data. First, the user can point the index finger to a position in space to get a reading of the data at that location. Second, cutting planes aligned with the principal planes can be obtained if the user forms a fist and moves it forward/backward, left/right, or up/down. Third, arbitrary cutting planes can be obtained if the user makes a flat posture with the hand as in a karate chop. An arbitrary cutting plane will then follow the position and orientation of the user's hand. Finally, if the user points with both the index and middle fingers, these fingers begin to act as a carving knife. The carving knife can be swept along a curve in space, generating a ruled surface. On this surface will appear a curved cross-section of the data set.

For these various forms of manipulation, the first thing that is needed is a way to recognize which hand posture the user is engaged in. To do this, an 18-value vector containing the joint angles from the glove is recorded. This vector is normalized to the "standard hand" by means of offsets and scales for each joint. This normalization ensures that for users with differently shaped hands, the normalized sensor readings for similar hand positions will have similar values. Next, the dot product is taken between this vector and each of the recognizable postures. The closest match is determined to be the current posture, unless the user's hand does not match any of the postures in which case the user is assumed to not be interacting with the data.

Once the current posture is determined, the system can activate the appropriate tool. For the flat cutting plane, the position and orientation of the plane are determined by measuring the 6D tracker attached to the user's wrist. From this, a plane can be placed in the correct location in the data set. The plane is triangulated adaptively so that the cross-sectional data values at the vertices of any triangle will not vary beyond some user-specified tolerance. Then the triangles are drawn using Gouraud shading hardware, assigning a color to each vertex from a color map that represents the range of values of the data.

The carving knife tool works similarly to the flat cutting plane, except that instead of drawing a single plane that follows the user's hand, a sequence of rectangles are drawn along the sweep of the hand. Each of the rectangles corresponds to the location of the virtual blade at a previous point in time. When the user moves his/her hand one blade's width away from the last rectangle, another rectangle is created to line up with the last. These rectangles are never removed until the user changes to another hand posture, at which time they all disappear.

Another good thing about letting the user manipulate which portion of the data set is of interest is that it prevents the program from having to compute and display the entire data set. In addition, this could also be used to cut down a huge data set into user-controlled local slices, allowing the user to interactively examine data that was previously too large and too slow to draw.

5 Results

To illustrate the results of the knife metaphor, we obtain planar and curved cuts from two different data sets. Each data set is a 20x20x20 regular grid. The first one contains a toroidal shape, while the second one contains a head shape. The shapes are of varying thickness and may contain holes. That is, the models do not use infinitely thin surfaces. Instead, they use walls that have thickness – that makes the head have a thickness yet have a hollow center. The grids also have distance values which refer to the distance from the walls. Yellow maps to zero distance (on either side of the wall), blue maps to negative distances (inside the wall), and red maps to positive distances (outside the wall – which may be inside or outside the object as a whole), with a smooth blend between colors.

6 Conclusion

We have presented some early results in our work on using common objects as metaphors for visualization. In particular, the use of knives to specify cutting planes have been extended to non-planar cuts. Aside from providing this extra degree of freedom in specifying regions of interest, we believe the work shows strong promise of these metaphors to aid the visualizer. Interaction is natural, and the user is not encumbered by having to remember awkward gestures; nor have to indirectly interact through intermediate handles.

References

[1] Gary Bishop et al. Research directions in virtual environments: Report of an NSF invitational workshop. In *Computer Graphics*, pages 153–177, August 1992.

[2] John M. Carroll and Mary Beth Rosson. Putting metaphors to work. In *Graphics Interface '94*, pages 112–119, 1994.

[3] Tom Meyer and Al Globus. Direct manipulation of isosurfaces and cutting planes in virtual environments. Technical report, Dept. of Computer Science, 1993. Technical Report 93-54.

[4] Robert C. Zeleznik et al. An interactive 3d toolkit for constructing 3d widgets. In *Computer Graphics*, pages 81–84, 1993.

[5] Henry Fuchs. Siggraph '91 Demo, 1991.

[6] Steve Bryson and Creon Levit. The Virtual Windtunnel: An environment for the exploration of three-dimensional unsteady flows. In *Proceedings: Visualization '91*, pages 17 – 24. IEEE Computer Society, 1991.

[7] David J. Sturman. *Whole-hand Input*. PhD thesis, Massachusetts Institute of Technology, 1992.

[8] Alex Pang, Naim Alper, Jeff Furman, and Jiahua Wang. Design issues of spray rendering. In *Proceedings: Compugraphics '93*, pages 58 – 67, 1993.

[9] Alex Pang. Spray rendering. *IEEE Computer Graphics and Applications*, 14(5):57 – 63, 1994.

[10] William Reeves. Particle systems: A technique for modelling a class of fuzzy objects. In *Computer Graphics*, pages 359–376, 1983.

[11] Craig Reynolds. Flocks, herds, and schools: A distributed behavioral model. In *Computer Graphics*, pages 25–34, 1987.

[12] Johan Stolk and Jarke J. van Wijk. Surface-particles for 3d flow visualization. In *Proceedings Second Eurographics Workshop on Visualization in Scientific Computing*, 1991.

[13] William Lorensen and Harvey Cline. Marching cubes: A high resolution 3D surface construction algorithm. In *Computer Graphics*, pages 163–169, 1987.

[14] J. P. M. Hultquist. Constructing stream surfaces in steady 3d vector fields. In *Proceedings: Visualization '92*, pages 171 – 177. IEEE Computer Society, 1992.

[15] Alex Pang and Naim Alper. Bump mapped vector fields. In *SPIE & IS&T Conference Proceedings on Electronic Imaging: Visual Data Exploration and Analysis*. SPIE, Feb 1995.

[16] Alex Pang, Craig M. Wittenbrink, and Tom Goodman. CSpray: A collaborative scientific visualization application. In *Proceedings SPIE IS&T's Conference Proceedings on Electronic Imaging: Multimedia Computing and Networking*, Feb 1995.

[17] Thomas Baudel and Michel Beaudouin-Lafon. Charade: Remote control of objects using free-hand gestures. *Communications of the ACM*, 36(7):28–35, July 1993.

[18] Colin Ware and Steven Osborne. Exploration and virtual camera control in virtual three dimensional environments. In *Computer Graphics*, pages 175–183, March 1990.

[19] Roger E. Axtell. *Gestures: The Do's and Taboos of Body Language Around the World*. John Wiley & Sons, 1991.

Editors' Note: see Appendix, p. 149 f. for coloured figures of this paper

Beyond volume rendering: visualization, haptic exploration, and physical modeling of voxel-based objects

Sarah F. Frisken Gibson

Mitsubishi Electric Research Laboratories, Cambridge Research Center
201 Broadway, Cambridge, MA, 02139

Abstract. This paper proposes the use of a voxel-based data representation not only for visualization, but also for physical modeling of objects and structures derived from volumetric data. Work in progress that demonstrates the utility of a voxel-based data format for modeling physical interactions between virtual objects is discussed, data structures that help to optimize storage requirements and preserve object integrity during object movement are presented, and prototype systems are described. These prototypes include 2D and 3D systems that illustrate voxel-based collision detection and avoidance, a force-feedback system that enables haptic, (or tactile), exploration of virtual objects, and a 2D system that illustrates interactive modeling of deformable voxel-based objects.

1 Introduction

For some data sources, a voxel-based representation can provide much more informative data visualization than the surface-based representation of conventional computer graphics. For example, the image data from a 3D Magnetic Resonance Image (MRI) scan consists of a wealth of information about internal structure, function, and anatomy. Using volume rendering with appropriate preprocessing, this information can be effectively presented to the attending physician or radiologist. In contrast, surface models derived from the MRI data discard internal structure, can not be used to represent objects with poorly defined edges or surfaces, and are sensitive to image segmentation errors.

In this paper, we propose the use of a voxel-based data representation not only for visualization, but also for physical modeling of objects and structures derived from volumetric data. The paper describes work in progress that demonstrates the utility of a voxel-based data format for modeling physical interactions between virtual objects. Data structures are presented that help to optimize storage requirements and preserve object integrity during object movement. An adaptation to volume rendering algorithms is discussed that enables objects to be rendered individually and then combined in a final compositing step. Finally, algorithms and prototype systems are presented that use a voxel-based format to model physical interactions between objects. These physical interactions include collision detection and avoidance, haptic, (or tactile), exploration of virtual objects using a force feedback device, and shape changes of deformable objects.

2 Motivation

Recent trends have provided a growing interest in using 3D images from medical imaging technologies such as MRI or Computed Tomography (CT) not only to visualize anatomy, but also to generate patient-specific anatomical models for pre-surgical planning and surgical simulation. Such models have been used for implant design in orthopedics, for planning surgical approach, for training and education, and for predicting the outcome of a planned surgery (for examples, see papers in [6, 20]).

The use of conventional graphics for object manipulation and physical modeling require that the 3D volumetric images be converted into surface-based or solid geometry-based models. However, such conventional-graphics models discard interior structure and are prone to estimation errors, especially where details in the surface (such as small fractures in a bone surface) are small relative to the resolution of the data set. Hence, just as volume rendering can be more powerful than surface-rendering methods for visualizing volumetric data, voxel-based object models which preserve interior structure may be more appropriate than surface-based models for physical modeling of objects derived from volumetric data.

A voxel-based graphics system that is capable of performing pre-surgical planning and surgical simulation must be capable of modeling the movement and interaction among objects in the scene. In most cases, objects must be prevented from interpenetrating, requiring that collisions between objects be detected and avoided. When collisions occur, virtual objects should respond in a physically realistic way so the graphics system must be able to model energy transfer between voxel-based objects and shape changes of deformable objects. In the following, several approaches and prototypes of a voxel-based graphics system are described that address these requirements.

3 Background

New imaging and data-sampling technologies have resulted in large, multi dimensional arrays of data. Conventional sources of volumetric data include: medical imaging technologies, such as MRI or CT; spatial data sampling, such as geological core sampling or the measurement of temperature or pressure by weather balloons for climate studies; and numerical simulations such as 4D studies of fluid flow or turbulence. More recent sources of volumetric data include industrial CT, finite-element analysis techniques, and "voxelization" methods used to convert CAD and CSG models into voxel-based data representations that can be visualized using volume rendering [11].

Early methods to visualize volumetric data used surface rendering techniques to display polygon-based surface models derived from the volumetric data. Since the late 1970's, volume rendering techniques have been developed which enable more direct visualization of the volumetric data. With appropriate preprocessing, volume rendering can be used to visualize surfaces, interior structure, and objects that do not have well-defined surfaces. Many effective volume-editing tools, volume rendering algorithms and data-compression schemes have been developed (see for example [13]. Although processing needs and large memory requirements still present major hurdles, faster algorithms and special-purpose hardware have enabled real-time volume rendering of data of significant size and resolution [4, 18, 19].

Recently, Kaufman et al [10] have introduced the field of volume graphics, where a voxel-based data format is used to represent graphical objects that are customarily represented by surface-based models. They have demonstrated that many graphical visual effects possible with surface-based graphics, including shading, reflectance, and radiosity, are also possible using volume graphics [2, 12, 23]. Although there are still many challenges to overcome, including large memory needs and fast data processing requirements, Kaufman et al. assert that volume graphics has the potential to supersede surface-based graphics just as 2D raster graphics superseded vector graphics. Whether or not this potential is realized will depend on many factors. However, as occurs in data visualization, some objects will be more accurately modeled using a voxel-based volume graphics format than conventional graphics formats.

4 Data Structures

There are three important considerations when designing data structures for interacting, voxel-based objects. First, the data structure of the virtual environment must be simple and easily updated. While many data structures have been devised to reduce memory requirements of voxel-based data, these generally require significant pre-processing and are not suitable for an environment in which objects can continuously move and change shape. Perhaps the simplest approach would be to represent the entire virtual environment by a (very large) discrete 3D array of voxels. As objects move about the space, the object's voxels would be shifted into new locations in the virtual environment. As we will see in section 6, this organization of the virtual environment leads to a very simple method for detecting collisions between objects.

The second consideration is that the amount of data for each voxel-based object can be very large. Not only is the 3D virtual environment large, but the data structures for individual voxels can be quite long. An individual voxel can consist of measurements of color, opacity, vector gradient, reflectance parameters, material strengths and other properties. While a large 3D voxel array of the entire virtual environment greatly simplifies modeling of physical interactions between objects, storing the lengthy voxel data structures directly in such an array is very inefficient when there is a lot of empty space between objects. As a consequence, voxel data for each object must be stored as compactly as possible while allowing fast data access for volume rendering and for updating the object position and shape.

A third important consideration is that, because of the discrete nature of the object representation, if the voxels are stored in a fixed, rectangular, data grid, then object voxels must be resampled for each non-integer movement of the object. The problem with this approach is that accumulated sampling error would quickly lead to complete degradation of a moving object. An optimal data representation would keep the individual object voxel data intact while somehow transforming the data grid in which the voxels lie.

13

Figure 1 illustrates the data structures used by algorithms and prototype systems discussed in this paper. Two separate data structures are used to minimize storage requirements and maintain the integrity of moving objects while providing an easily updated structure for modeling physical interactions among voxel-based objects in the virtual environment. An occupancy map with dimensions of the virtual environment is used to monitor interactions among objects and global changes in the environment. Each element of the occupancy map is large enough for a single address and contains either a null pointer or an address of an object voxel data structure. Voxel data structures consisting of elements such as color, opacity, reflectance coefficients, vector gradient, material properties, position, neighbor addresses, etc. are stored in compact object arrays. The occupancy map is initialized by mapping the addresses of voxels in each object into the occupancy map using the voxel positions in object-space coordinates and a mapping transformation from object space to the virtual-environment space.

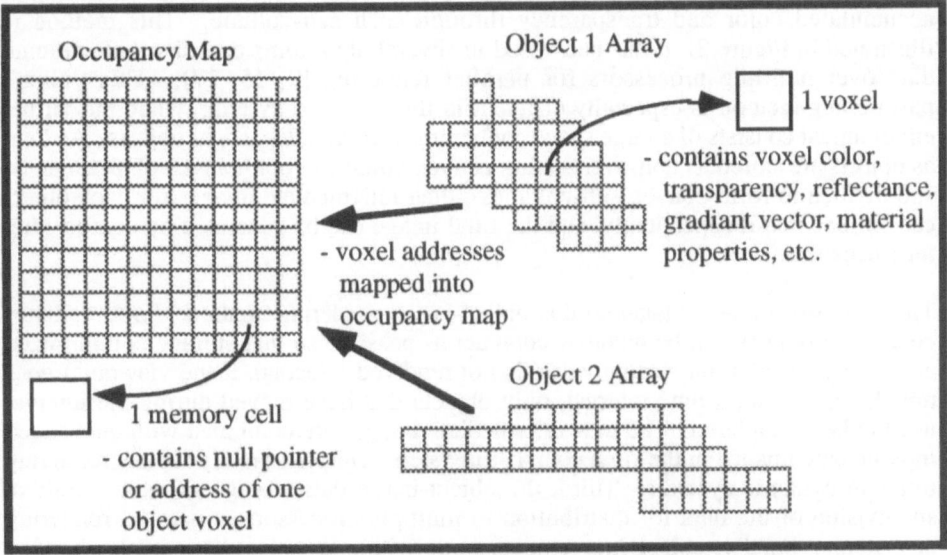

Fig. 1. Data Structures for a voxel-based object representation. The occupancy map consists of regularly spaced memory cells of integer size that contain either a null pointer or an address of one object voxel. The occupancy map is the size of the application's virtual world. The object arrays consist of data structures each containing the information for a single voxel. This voxel data structure can include many properties and attributes including voxel color, transparency, gradient vector, position vector, material properties, etc. Object arrays are not necessarily rectangular; object voxels are not necessarily evenly spaced or of equal sizes; and voxels are not necessarily ordered within the array. Since a voxel-based graphics system is very memory intensive, the purpose of the double data structure is to pack the large voxel structures of the object as efficiently as possible while maintaining a map of the entire virtual space for modeling interactions between objects.

5 Object-based Volume Rendering

In the proposed data structures, the data required for volume rendering is stored in the object arrays. However, in most volume rendering algorithms (for semi-transparent or opaque data), the entire volume to be rendered is traversed in order, either from the front of the data volume (facing the viewing plane) to the back of the data or from back-to-front. Using the proposed double data structure, this method would require traversing the 3D occupancy map in order and accessing the voxel data at the address in occupied cells intersected by the rendering rays. Because of limits in memory page sizes and available RAM, this address lookup from occupancy map to object array is an extremely inefficient way to access the voxel data and would greatly limit the rendering speed.

Fortunately, a large data volume segmented into non-overlapping convex sub-volumes can be rendered by calculating and compositing 2D intermediate images consisting of accumulated color and transparency through each sub-volume. This method is illustrated in Figure 2. It has been used in several algorithms that distribute volume data over multiple processors for parallel rendering [9, 15, 17]. This volume partitioning method is especially efficient in the proposed system, where the virtual environment consists of a large space containing several independent objects. As long as objects are stored as non-overlapping convex volumes, (non-convex objects can be sub-divided to form convex sub-volumes), then intermediate images for each object can be rendered independently, and the final image can be generated by compositing intermediate images

There are several advantages to this object-based rendering method. First, volumes containing objects can be made as compact as possible so that storage is minimized and empty space in the virtual world is not rendered. Second, if the viewpoint does not change during a time interval, only objects that have moved during the interval need to be re-rendered. The new intermediate images are combined with unchanged intermediate images in the final compositing step. This can greatly reduce rendering times in dynamic systems. Third, the object-based data storage gives an intuitive subdivision of the data for distribution to multiple processors in parallel rendering algorithms. Finally, object data remains static within an object array even when the object moves about the virtual environment. Movement of an object relative to the virtual environment simply requires an update to the coordinate mapping transformation between the object and the environment. Objects are not resampled when they move, so that object degradation due to accumulated resampling errors is avoided.

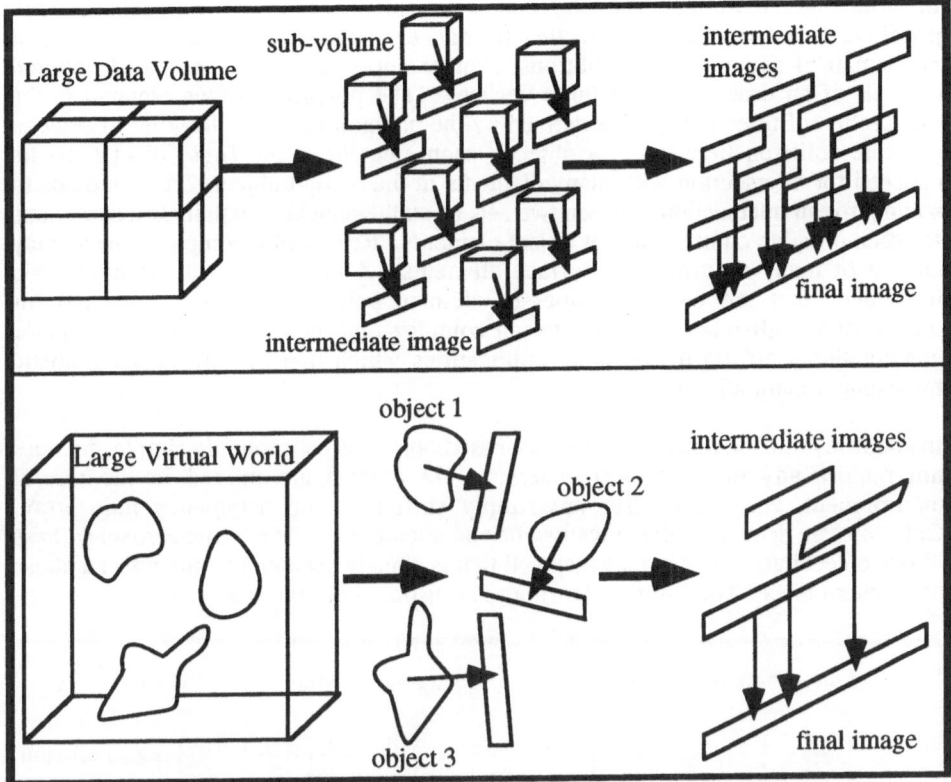

Fig. 2. Partitioning of the data for parallel volume rendering. For a large, regular volume, the volume is partitioned into cubes. Each cube of data is rendered separately, forming a 2D intermediate image that contains accumulated color and opacity. These intermediate images are then composited to form the final image. For the virtual environment, individual object data are kept in static arrays. Each time an object is moved or the viewpoint changes, the object's intermediate image is re-rendered using the new object-space to image-space transfer function. Whenever an object's intermediate image is re-rendered, the final image is recomposited and the display is updated.

6 Collision Detection

For realistic modeling of objects in a virtual environment, it is important to model interactions between objects in a physically realistic way. When virtual objects come in contact, they must be able to exchange momentum and energy, in many cases they must be prevented from penetrating each other, and they may deform on impact. The first step of modeling interactions between objects requires the detection of collisions between moving objects.

In surface-based graphical formats, objects are modeled as lists of polygons or other primitive surfaces elements. In this format, each element is essentially a set of mathematical equations or conditions. For example, a single three-sided polygon consists of an ordered list of three vertices. The polygon surface element is the intersection of three half planes defined by the polygon vertices. In order to detect a possible collision between two objects, each element in the first object must be checked for intersection with many elements in the second object. This amounts to solving for an intersection between two sets of mathematical equations and conditions for each possible combination of object elements. Reasonably complex objects may consist of many thousands of surface elements. Although algorithms have been introduced that reduce the number of element pairs that must be checked for intersection, collision detection between complex objects remains one of the major computational efforts in computer applications which perform physically realistic modeling or simulation [3].

In contrast, collision detection for voxel-based objects is conceptually simple and does not require any mathematical analysis. As objects are moved in the virtual environment, voxel addresses are simply shifted in the occupancy map array. Collisions are detected automatically when an attempt is made to write a voxel address of one object into an occupancy map cell that is already occupied by the voxel address of another object. This simple algorithm is illustrated in Figure 3.

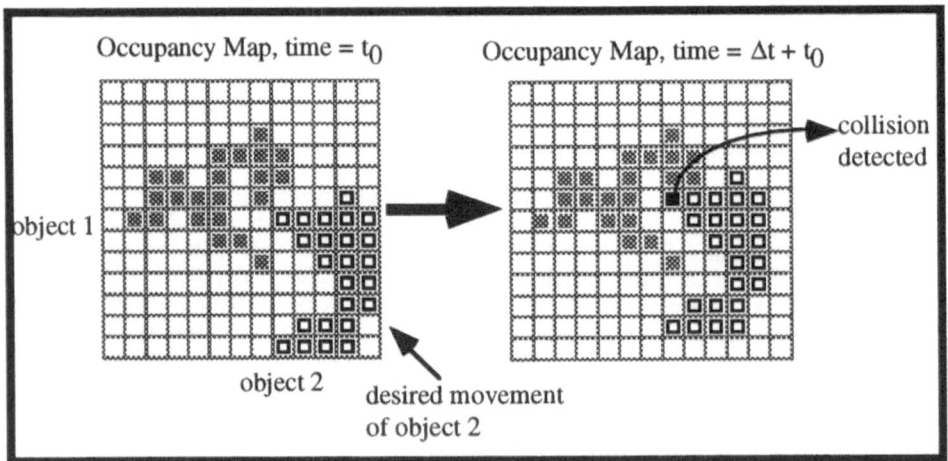

Fig. 3. Collision detection algorithm. As object 2 moves along its desired trajectory, the addresses of the object are shifted in the occupancy map. If an attempt is made to shift an object 2 address into an occupancy map cell that is already occupied by an address of object 1, then a collision is detected.

The data structures that have been proposed here suggest the following simple algorithm for collision detection and avoidance:

```
determine desired object position
while (stepsize to desired position > threshold) {
        for (all voxels in the moving object) {
                check occupancy map cells of desired positions to see if they
                        are occupied
        }
        if (all occupancy map cells are free) {
                move the object into the desired position
                update the occupancy map
                exit
        }
        else if (any of the occupancy map cells is occupied) {
                avoid collision by reducing the step size
                update the desired position
        }
}
```

7 Collision-Detection and Avoidance Prototypes

Two prototype systems have been built to test this collision detection and collision avoidance algorithm. In a 2D prototype system, the occupancy map is updated as the positions of 2D objects are manipulated with the computer mouse. Applications that have been developed for this prototype include a virtual maze where objects with complex edges are moved between the voxel-based maze walls. Collisions with the maze walls are detected and the objects are restricted to the free space between the maze walls. In a second application, 2D virtual puzzle pieces with arbitrary edges can be cut from a digitized color photograph and assembled using a computer mouse.

In a 3D prototype system, voxel-based nut, washer, and bolt objects are interactively manipulated, visualized and assembled. The object positions and orientations are controlled interactively with a 6 degree-of-freedom Fastrack 3Ball input device by Polhemmus. Computation and rendering was performed on an SGI Reality Engine although only minimal effort was made to optimize code or to take advantage of specialized hardware. Memory limits and the need for interactive rates limited the sizes of object volumes to less than about 680 kbytes. Volume rendering quality was compromised in order to attain interactive speeds (updates of 3 to 10 frames per second) but the rendering quality was sufficient for good visualization of object position.

Using this 3D prototype system, objects can be moved, visualized and collided with walls and other objects at interactive rates. With some dexterity, it is possible to assemble the washer on the bolt. However, poor object position and orientation control and lack of force feedback in the current system makes it too difficult to thread the nut on the bolt. Planned improvements to the system include the addition of depth perception for better control of object placement along the z-axis, calculation of collision forces for simulating more physically realistic reactions to collisions, and force feedback for better perception of collisions.

8 Haptic Exploration of Large Data Volumes

One direct application of the voxel-based collision detection and avoidance algorithms is a system that uses force feedback to enable the user to haptically explore or "feel" the 3D surface of a virtual object. Haptic, or tactile, exploration can enhance visualization methods for complex data and provide valuable feedback in the simulation of deformable objects. Using the proposed data structures along with a relatively small voxel-based model of the user's hand or fingertip, the computation required to detect avoid collisions and avoid penetration of the virtual object by the hand model is minimal even when the data itself is very large and complex.

A prototype system, illustrated in Figure 4, is currently under development for haptic exploration of a high resolution 3D CT scan of the pelvic bones of a human hip. The hip data is stored in a large static array. For haptic exploration, the data for each voxel consists of tissue density, measured direction from the CT intensity data, and a surface normal vector, calculated from the CT data. A force-feedback system [16] is used to track the user's fingertip position in the virtual environment. A small voxel-based object representing the user's hand or fingertip is shifted through the data array, tracking movement of the user's hand. When the fingertip model encounters a voxel belonging to the hip surface, the force-feedback device prevents the user's fingertip from moving in a direction that would penetrate the virtual object. Using the voxel's surface normal vector, the user's hand is restricted to movements parallel to or away from the virtual surface of the pelvic bones, allowing the user to "feel" the bone surface.

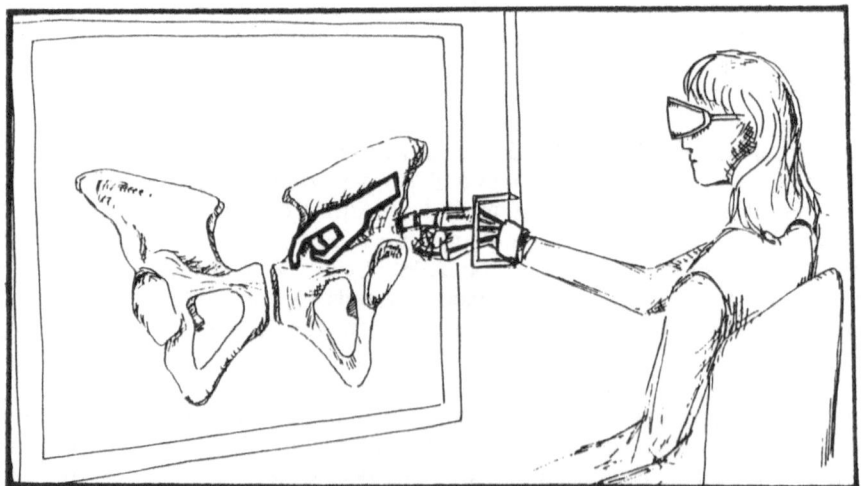

Fig. 4. Illustrates a prototype system for haptic exploration of a virtual object using a force-feedback device. Voxel-based collision detection and avoidance algorithms are used as input to a force feedback device to limit the user's hand position as it explores the data volume. By restricting hand motion so that it can not penetrate the virtual model, the user is able to "feel" the voxel-based virtual object.

Because of limits in volume rendering speed in the prototype system, user controlled movement of the hip model is limited to rotation about the vertical axis. High-quality volume-rendered stereo image pairs of the hip are pre-computed and stored for fast access. These are composited with a dynamically-computed rendering of the fingertip model and presented to the user via stereo goggles. This method gives some control of the view point to the user while providing real-time, high quality volume rendered images for visualization.

9 Object Deformation

A system that models physically realistic interactions must be able to model both rigid and deformable objects. In surgical simulation, we would like to be able to model deformation and tearing of skin during an incision, stretching or contraction of a muscle during limb movement, or the deformation of an organ when it is probed by a surgeon's instrument. Some very interesting work has been done in modeling skin using finite-element methods [20, 22]. However, finite-element methods are computationally expensive because they require solutions of large systems of differential equations. In interactive, real-time applications, only small numbers of finite elements may be used.

In a voxel-based object model, the number of elements in the deformable model is equal to the (potentially large) number of voxels in the object. Voxels in the object are joined to their nearest neighbors by a set of spring-like connections as shown in figure 5a). The connections act like non-linear springs which maintain the structure of the object, tending to pull the voxel mesh back towards an equilibrium state after the object is deformed by external forces. Because the number of interconnections between voxels is large, traditional finite element methods to determine final voxel positions at each time step are impractical. Instead, voxel positions are adjusted at each time step by the amount dependent only on distances to the nearest neighbors and independently of adjustments to neighbors. A closed feedback locp that locally adjusts each position, itteratively pulls each object towards an equilibrium state. If the feedback system is stable, this object will eventually settle in an equilibrium state after a finite disturbance. Figure 5b), illustrates a voxel that is pulled towards an equilibrium position equidistant between its four nearest neighbors.

The data structure used to model deformable objects is essentially the same as that described above and illustrated in Figure 1. A double structure is used, with object interactions modeled in a regular, discrete, occupancy map, and object voxel data stored in a compact object array. An additional consideration related to the issue discussed earlier of avoiding object degradation with non-integer object movements must also be addressed. If the object is stored in a regular, discrete array, then object voxels must be resampled when the object is deformed. The sampling errors introduced by a sequence of deformations would quickly degrade the object. In order to avoid this problem, floating point (x, y, z) positions of each voxel are stored as part of each voxel structure. Whenever the object is moved or deformed, these voxel positions are updated in the object array.

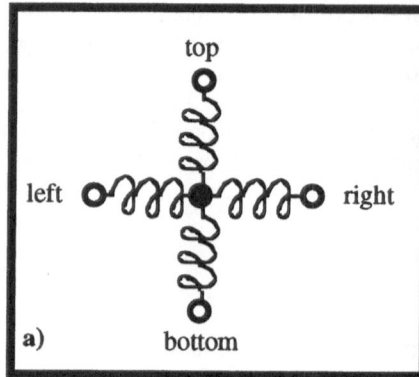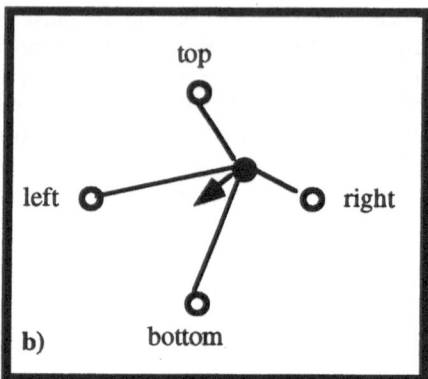

Fig. 5. a) Illustrates the non-linear springs connecting each 2D voxel to its four nearest neighbors. **b)** At each time step, the distances from the current voxel (dark circle) to its nearest neighbors (open circles) are examined. If these distances do not satisfy an equilibrium condition, the current voxel is moved closer to an equilibrium position. In this example, the current voxel is moved towards the point equidistant from its nearest neighbor.

Although this strategy solves the problem of object integrity during object movement and deformation, the data can now lie in a non-rectangular, non-uniform grid. Each voxel can have a unique shape and its position relative to its neighbors in not deterministic. Since existing algorithms and software for fast volume rendering depend on data coherence, current interactive rendering approaches can not be used. For this reason, we have had to limit prototypes of deformable voxel-based systems to 2D. We hope to look at hardware approaches to enable interactive volume rendering of such irregular data volumes in the future.

There are three basic object types that must be modeled: rigid objects, whose shapes and inter-voxel spacing cannot change; elastic objects, whose shapes can change within some allowable range but which will settle back to their original form when external forces are removed; and deformable objects, whose shapes can change permanently when forces are applied.

In order to model these three object types, we define two regions about each object voxel, the allowable region, and the deformation region. The size and shape of these regions define how an object will react to external forces. Figure 6 illustrates one of these regions in a 2D system. When an object is initially subjected to a force, each voxel in the object shifts the minimal amount possible to a position within its allowable region relative to its nearest neighbors. In a rigid system the allowable region has zero area and all object voxels are moved by the same amount when the object is subject to an external force. In an elastic or deformable system, the allowable region has non-zero area and the object voxel positions can move relative to each other to satisfy the force constraints. After the initial force application, the system itteratively adjusts voxel positions until an equilibrium state is reached (or until a new force is applied). This equilibrium state is determined by the deformation region since each voxel is adjusted until it lies within the deformation region relative to its nearest neighbors. In an elastic object, the area of the deformation region is zero and the equilibrium state is the original inter-voxel spacing. In a deformable object, the size of the deformation region defines how deformable, or plastic, the object is. Table 1 illustrates the relationship between object type and the areas of the allowable and deformation regions.

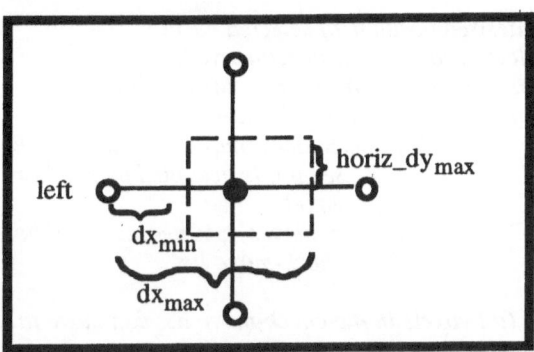

Fig. 6. Illustrates the allowable or deformation region of the current voxel (dark circle) with respect to its left neighbor. If this is the allowable region, then as long as the external force does not require the current voxel to move out of this region, the left voxel will not be required to move. If this is the deformation region, then as long as the voxel lies within this region, the current voxel is at equilibrium with respect to its left neighbor. The size of the allowable region affects how much the object can be stretched. The size of the deformable region affects how much the object can be defromed.

	rigid object	elastic object	deformable object
allowable region	zero area	non-zero area	non-zero area
deformable region	zero area	zero area	non-zero area

Table 1. Illustrates the relationship between object type and the areas of the allowable and deformable regions. If the allowable region has zero area, object voxels must remain at a fixed position relative to their nearest neighbors when the object is moved. When the area of the deformation region is zero, the equilibrium state is reached when all voxels lie at their original position relative to their nearest neighbors.

Note that, for both the allowable and deformation regions, the region shape can be different along different axes of the object. For example, an object could be rigid along one axis and deformable along another. This flexibility is important for modeling anisotropic materials: materials with different material properties along different axes. For example, skeletal muscle can contract and stretch much more easily along its longitudinal axis than along its transverse axis.

The following algorithm can be used to model object movement and deformation. Note that the same algorithm is used to model rigid, elastic, and deformable objects. The algorithm assumes that the force is applied to the selected voxel.

```
determine desired position of selected voxel
add the selected voxel to the check-move list
while (stepsize to desired position > threshold) {
        for (next voxel in the check-move list ){
                check occupancy map cell of nearest allowable position to
                        see if it is occupied by another object or a fixed
                        voxel
                if the voxel moves, add nearest neighbors to the
                        check-move list
        }
        if (all voxels in the check-move list can move to their allowable
                region) {
                move the voxels into the allowed positions
                update the occupancy map
                exit
        }
        else if (any voxel in the check-move list cannot move to its
                allowable region) {
                reduce the step size
                update the desired position of the selected voxel
        }
}

while (not all object voxels are at equilibrium) {
        locally adjust each voxel position towards its deformation region
}
```

A 2D prototype has been built to test algorithms for moving and deforming a single voxel-based object. Using a computer mouse, any voxel in the object can be grabbed and pulled, causing the entire object to deform and move subject to the user defined allowable and deformation regions. By varying the size and shape of these regions, very complex objects and materials can be interactively modeled. In this prototype, the deformable object can be collided with a wall or wrapped around a fixed object. Future work will include modeling interaction between two or more deformable objects, improving algorithms and rendering speeds, and extending this system to 3D.

10 Summary

This paper has introduced the use of a voxel-based data format to model 3D, volumetric, deformable objects in a virtual environment. Several prototype systems have been developed to demonstrate some of the potential of a voxel-based system. These prototypes include 2D and 3D collision detection, a 3D force feedback system for haptic exploration of volume data, and a 2D voxel-based deformable object. There are many challenges for developing a large interactive virtual environment using voxel-based objects, including limits due to rendering speeds, data-access speeds, and memory requirements. However, these challenges are no greater than those originally faced by conventional graphics. Further development of intelligent data structures, fast, parallel algorithms, and special-purpose hardware will enable the exploration and realization of the full potential of volume graphics.

References

1. S. Arridge, "Manipulation of Volume data for Surgical Simulation", in "3D Imaging in Medicine", eds. K. Hohne et al, Springer-Verlag, 1990.

2. Avila, R., Sobierajski, L., Kaufman, A., "Towards a Comprehensive Volume Visualization System", Proceedings, IEEE Visualization '92, pp. 13-20, 1992.

3. Baraff, D., "Rigid Body Simulation" in "An Introduction to Physically-Based Modeling", Course Notes 32, organizer: A. Witkin, Siggraph, 1994.

4. B. Cabral, N. Cam, J. Foran, "Accelerated Volume Rendering and Tomographic Reconstruction Using Texture Mapping Hardware", Proc. IEEE 1994 Workshop on Volume Visualization, Washington, DC, pp. 91-97, 1994.

5. L. Chen and M. Sontag, "Representation, Display, and Manipulation of 3D digital scenes and their Medical Applications", Computer Vision, Graphics, and Image Processing, 48, 1989, pp. 190-216.

6. A. DiGioia, T. Kanade, R. Taylor, eds."Proceedings of the First International Symposium on Medical Robotics and Computer Assisted Surgery", Shadyside Hospital, Pittsburgh, PA, 1994.

7. J. Gerber et al., "Simulating Femoral Repositioning with Three-dimensional CT", J. Computer Assisted Tomography, 15, 1991, pp. 121-125.

8. Hohne, K.H. et al, "3D-Visualization of Tomographic Volume Data Using the Generalized Voxel Model", The Visual Computer, 6, February, 1990, pp. 28-37.

9. Hsu, W., "Segmented Ray Casting for data Parallel Volume Rendering", Proc. 1993 Parallel Rendering Symposium, ACM Press, 1993.

10. Kaufman, A., Cohen, D., Yagel, R., "Volume Graphics", Computer, 27, July 1993, pp. 51-64.

11. Kaufman, A., "Efficient Algorithms for 3D Scan-Conversion of parametric Curves, Surfaces, and Volumes", Computer Graphics, 21, July 1987, pp. 171-179.

12. Kaufman, A. and Bakalash, R., "Memory and Processing Architecture for 3D Voxel-Based Imagery", IEEE Computer Graphics and Applications, pp. 10-23, 1988.

13. Kaufman, A.,ed., "Volume Visualization", IEEE CS Press, Los Alamitos, CA, 1990.

14. Kaufman, A., Yagel, R., Cohen, D., "Intermixing Surface and Volume Rendering", in *3D Imaging in Medicine: Algorithms, Systems, Applications*, K.H. Hoehne, H. Fuchs, and S.M. Pizer, eds., Springer-Verlag, Berlin, 1990, pp. 217-227.

24

15. Ma, K-L, Painter, J., Hunsen, C., Krogh, M., "A Data-distributed Parallel Algorithm for Ray-traced Volume Rendering", Proc. 1993 Parallel Rendering Symposium, ACM Press, 1993.
16. T. Massie, K. Salisbury, "The PHANToM Haptic Interface: A Device for Probing Virtual Objects", Proc. ASME Symposium on Haptic Interfaces for Virtual Environments and Teleoperator Systems", Chicago, Nov. 1994.
17. Neumann, U., "Volume Reconstruction and Parallel Rendering Algorithms: A Comparative Analysis", Ph.D. dissertation, Dept. Computer Science, U.N.C. Chapel Hill, 1993.
18. Neumann, U., "Parallel Volume Rendering Algorithm Performance on Mesh-Connected Multicomputers", Proc. 1993 Parallel Rendering Symposium, ACM Press, 1993.
19. Pfsiter, H., Kaufman, A., and Chiueh, T., "Cube-3: AReal-Time Architecture for High-Resolution Volume Visualization", Proc. 1994 Workshop on Volume Visulization, Washington, DC, pp. 75-83, 1994.
20. Pieper, S., Rosen, J., Zeltzer, D., "Interactive Graphics for Plastic Surgery: A Task-level Analysis and Implementation", ACM Proc. Interactive 3D Graphics, 3, pp. 127-134, 1992.
21. R. Robb, ed., "Visualization in Biomedical Computing 1994", SPIE 2359, 1994.
22. Terzopoulos, D., Waters, K., "Physically-based Facial Modelling, Analysis, and Animation", J. Visualization and Comp. Animation, 1, pp. 73-80, 1990.
23. Westover, L., "Footprint Evaluation for Volume Rendering", Computer Graphics, 24, August, 1990, pp. 367-376.
24. Yagel, R., "Realistic Display of Volumes", Image Capture, Formatting and Display, SPIE Vol. 1653, pp. 470-476.
25. Yasuda et al, "Computer System for Craniofacial Surgical Planning based on CT Images", IEEE Trans. on Med. Imaging, 9, 1990, pp. 270-280.

On–line Visualization of Arbitrary Unstructured, Adaptive Grids

T.Schmidt and R.Rühle

Computing Center (RUS), University of Stuttgart,
Allmandring 30, 70550 Stuttgart, Germany

Abstract. Computational simulations for complex geometries nowadays often deal with unstructured grids. That's why scientific visualization must also be extended to these grid types. A visualization system
is presented which works with any type of unstructured grids, i.e. arbitrary polygonal and polyhedral grids. Results from finite volume (cell–
centered) and finite element methods can be visualized. Additionally
time–dependent data can be processed and interpolation between time
steps is possible. This data may be transfered directly from a simulation
program and therefore on–line visualized. To integrate this task into a
computational simulation an object–oriented interface is presented which
allows an easy connection of simulation and visualization.

1 Introduction

Former scientific computing methods only worked with structured grids because
these are easier to deal with. But more and more software packages tend to use
unstructured grids. These grids may consist of a set of predefined polygonal or
polyhedral types. Some of them even use arbitrary unstructured grids. In [11] a
finite volume method is described; a finite element method is presented in [10].

A two–dimensional arbitrary unstructured grid is defined as a set of arbitrary
shaped polygons which build the computational domain. Hence neighbouring
polygons have common edges. In 3D the grid consists of polyhedrals. Analogous
neighbouring polyhedrals must share polygons. Figure 1 shows an example of a
two dimensional arbitrary unstructured grid. This could be used to compute the
flow around an octagon. It consists of four heptagons and two rectangles.

Some of the advantages of arbitrary unstructured grids as presented in [7]:

- automatic and easier grid generation
- grid generation may be integrated into the simulation process
- no additional effort for complex geometries
- no grid degenerations in critical regions

Therefore research in the field of visualization must be extended to arbitrary
unstructured grids which is the main topic of this paper. This work is closely

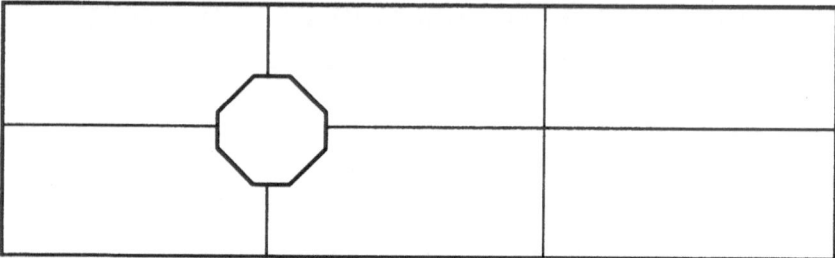

Fig. 1. Example of a 2 dimensional arbitrary unstructured grid

related with the work of Helf [7]. So in a first step the visualization system was developed for the needs of visualizing fluid flows.

The system is designed to handle results of finite element methods as well as of cell–centered finite volume methods. Additionally it is possible to define piece-wise continuous functions on the problem domain, i.e. continuous within polygons respectively polyhedrals and discontinuous on the cell boundaries.

One of the presented system's purposes is to help verifying numerical simulation packages during the development process.

The following sections describe the object–oriented data model and the architecture of the visualization system. A look on other visualization systems and further developments is done.

2 Data Model

Unlike to structured grid simulations the data model for the results of arbitrary unstructured grid methods is much more complex. This object–oriented model consists of the description of the complex geometry and a connection between geometry and solution values from the simulation depending on the used numerical method. Additionally time–dependency is covered by this model.

2.1 Grid and Solution Representation

The geometry is recursively defined. First, a node n_i is a vector of dimension dim with coordinates and at the same time an element of the lowest hierarchy level H_0.

$$n_i = (c_1, \ldots, c_{dim})$$
$$H_0 = \{n_1, \ldots, n_{nodes}\} \tag{1}$$

Further hierarchy levels H_i are defined as a set of subsets M_k of the previous level.

$$M_k = \{m_1, \ldots, m_l\} \text{ with } m_j \in H_{i-1}$$
$$H_i = \{M_1, \ldots, M_{elements}\} \tag{2}$$

A n-dimensional grid G is defined by a tuple of $n+1$ different hierarchy levels H_i.

$$G = (H_0, \ldots, H_n) \tag{3}$$

The solution values are yet an arbitrary number of scalars and vectors. On these solution values any number of evaluation functions can be defined. If the numerical simulation is a cell–centered finite volume method all solution values are assigned to the centers of the elements of the highest hierarchy level. The value v at a point x within a top level element is defined by a function f of the value v_0 at the cell center x_0 and the point's position.

$$v = f(v_0, x_0, x) \tag{4}$$

If the simulation uses a finite element method the solution values v_i are connected to the nodes of the grid. On all nodes which build a cell, element functions e_i are given. These functions may be discontinuous. Then the value v at any point x inside a cell with n nodes can be computed using the vectors x_i.

$$v(x) = \sum_{i=1}^{n} e_i(x - x_i)\, v_i \quad \text{with} \quad \begin{cases} e_i(x) = 0 & : \quad x = x_j, j \neq i \\ e_i(x) = 1 & : \quad x = x_j, j = i \\ 0 \leq e_i(x) < 1 & : \quad otherwise \end{cases}$$
$$\text{and} \quad \sum_{i=1}^{n} e_i = 1 \tag{5}$$

Consequently the whole data model is dimension independent. Figure 2 shows the inheritance tree for the data model. A hierarchy level H_i as defined in equation 2 which contains links to a lower hierarchy level, is called *connectivity*.

Some restrictions have to be made to ensure that the represented grid is valid. This means that the top level elements must be a decomposition of the simulation domain. All elements of all levels should be used.

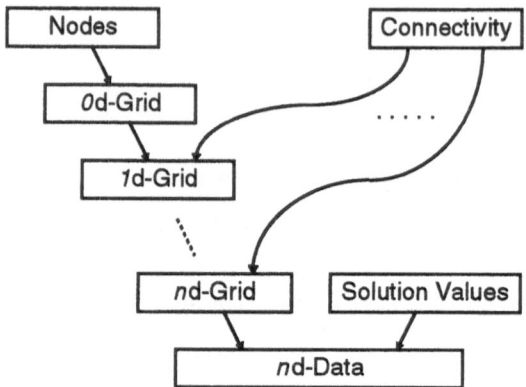

Fig. 2. Inheritance tree for the object–oriented data model

2.2 Representation of Time

Time–dependency is realized through a sequence of time steps. A time step T_i consists of a grid G_k, corresponding solution values S_i and an entity t_l which represents the time.

$$T_i = (G_k, S_i, t_l) \tag{6}$$

The complete data set which must be visualized is a set T of time steps T_i.

$$T = \{T_0, \ldots, T_{n-1}\} \tag{7}$$

These are sorted according to the time entity mentioned above.

Each time step has its own solution values, but several time steps may share the same grid. To allow an interpolation of time without any gaps the numerical simulation must provide two time steps at the time of a refinement. The solution is continuous at the refinement time if the refinement process is exact, i.e. the solution value of a parent cell which is refined, must be identical with the values of the child cells.

At time t_1 the data is interpolated between time steps b and c with identical grids of refinement level 1. The gird is refined at time $t_c = t_{c_r}$. Thus, at time t_2 the data is interpolated between time steps c_r and d. These time steps have grid data of refinement level 2. See figure 3 for a detailed view.

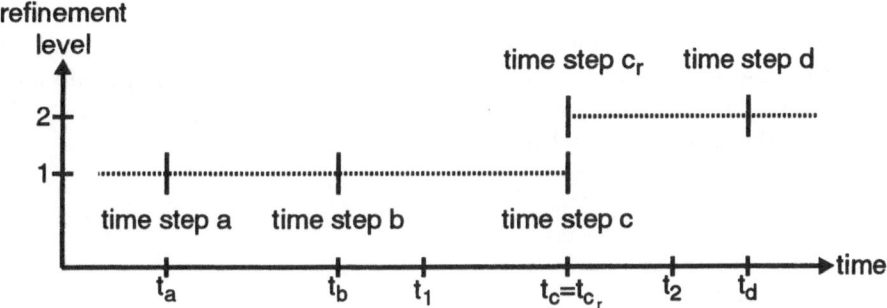

Fig. 3. An example time step list demonstrating linear interpolation

3   System Architecture

The whole system consists of three components. The first is the simulation package which shall be connected to the visualization system. The other two components are the visualizer and the renderer. The connection between the simulation package and the visualization system is done with an interface library, the *heterogenous simulation data interface* (HSDI). Figure 4 shows a system overview.

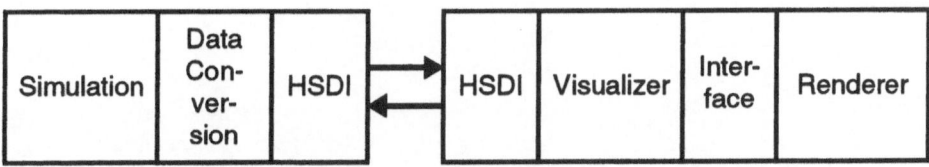

Fig. 4. Overview of the on–line visualization system

3.1   Heterogenous Simulation Data Interface

The *heterogenous simulation data interface* (HSDI) is an object–oriented library which was developed in C++. It provides methods for data conversion, transfer and file I/O. Using HSDI makes it possible to send data from the simulation directly to the visualization. Otherwise this data can be archived first and visualized later. For the on–line visualization two modes exist. The first one is the *tracking* mode. The simulation may send data at any time if the visualization is ready to receive. The second mode is the *independent* mode. Only if the visualization demands data it is sent by the numerical simulation. The necessary methods to manage these modes exist in the HSDI–library. The C++ –like pseudo code segments in figures 5 and 6 show how this can easily be implemented by the developer of the simulation package.

$$\vdots$$

```
HSDI.init();
// repeat the loop until the solution is good enough
do {
    // here the numerical code must be placed
    numerical.iteration();
    // the simulation may not send all iterations
    if (send_data_from_sim_to_vis == TRUE) {
        if (HSDI.readyToReceive() == TRUE) {
            numerical_data.convert(HSDIdata);
            HSDI.send(HSDIdata);
        }
    }
} while (abort_condition == FALSE);
```

$$\vdots$$

Fig. 5. Pseudo code to integrate on–line visualization (tracking mode)

$$\vdots$$

```
HSDI.init();
// repeat the loop until the solution is good enough
do {
    // here the numerical code must be placed
    numerical.iteration();
    if (HSDI.requestReceived() == TRUE) {
        numerical_data.convert(HSDIdata);
        HSDI.send(HSDIdata);
    }
} while (abort_condition == FALSE);
```

$$\vdots$$

Fig. 6. Pseudo code to integrate on–line visualization (independent mode)

Only a few lines must be inserted into the simulation code. Adaptive grid management considering time–dependent data is not presented in this short example. Few additional effort is needed for the tracking mode. The *independent* mode produces time gaps in case of grid update (refinement or re–coarsening) and is therefore more suitable to examine single time steps. Most work must be done for data conversion. But the HSDI–library supports this task by appropriate methods.

Further it is possible to send information from the visualization back to the simulation. The idea is to interactively control the simulation process. Using the HSDI named objects can be placed in the renderer. The user can select one of these objects and the name of the chosen object is transfered back to the numerical simulation. This feature can be used e.g. to intervene in the refinement process. Additional control commands can be sent by the visualization. The simulation code may ignore them or execute commands like restart, terminate or change parameters.

3.2 Visualization System

The data objects of the visualization are derived from the ones of the HSDI–library. The visualization data is organized in a list of time steps as mentioned in the previous section. At the moment only linear interpolation of time is implemented. Other interpolation methods are suitable for animations. So the time derivation of the solution can be continuous as long as the underlying grid does not change. On this way smooth animations are possible without the need of an enormous amount of time steps.

The visualization of a definite time step at first needs an interpolation method within a cell. For solution values of cell–centered finite–volume methods by default the value at the cell center is taken for the whole cell area. Hence the values inside the cell is constant. Another interpolation method is linear reconstruction using gradients provided by the simulations solution values. Other interpolation methods may be defined by the developer of the numerical simulation. The solution is continuous within a cell but may be discontinuous on the borders of the cells.

Results of a finite–element method are computed in another way. On all nodes which build a cell, element functions can be specified by the developer of the finite element code. These functions may also be discontinuous. The values within a cell can be computed by using the node values and the assigned element functions.

On this way a complete solution for the domain given by the grid is computed. Selected solution values may be displayed by color (2D and 3D) or height encoding (2D). The advantages of an object–oriented model are used for the realization of further visualization operators similar to the suggestions in [4]. These are given by dimension independent methods. The *iso*–method returns isolines for 2D and isosurfaces for 3D. Also stream lines and particle tracing will be implemented.

Polygon lists generated by the visualization methods are then moved to the renderer. Open Inventor™, an object–oriented 3D toolkit developed by Silicon Graphics, was used for the implementation of the renderer. This toolkit is supported by several workstation manufacturers. An example of an on–line visualization of the CFD–code from [7] is shown in figure 7 (see Appendix). It shows the time–dependent simulation of an internal supersonic flow around a circle using adaptive grid refinement. The left column shows the grid on the different refinement levels. From the third time step a detail of the grid is shown. In the middle column color–encoded energy is presented. The right column shows the height–encocded density values. Here the gradients calculated by the simulation can be seen.

4   Comparisons with Other Visualization Systems

The main difference between the presented visualization system and other systems is the management of arbitrary polygonal and polyhedral grids. Piece–wise continuous functions and results of cell–centered methods can also be visualized.

Commercially available visualization systems deal with limited unstructured grids. Within IRIS Explorer™ for example only a limited number of basic cells (called dictionary elements) is available. In 3D these are tetrahedron, pyramid, prism, deformed brick (wedge) and brick [6]. Grids constructed using these elements are useful to visualize finite element–based simulations. The visualization of finite volume, cell–centered simulation is not possible. Also the data is usually read from files and no on–line facilities are given. Unsteady simulation data cannot be processed either.

GRAPE (GRAphics Programming Environment) is a flexible tool for the visualization of finite elements on triangular and tetrahedral grids [5]. The handling of unsteady simulations is integrated. Management of time–dependent data as described in [9] influenced the presented work and lead to a similar model. The possibililty of on–line visualization is not integrated.

The Cortex visualization system [2] is designed for interactive analysis and display of CFD simulation data. These solvers can be based on triangular and quadrilateral cells (2D) or tetrahedral and hexahedral cells (3D). Examples are given in [1]. In addition to the features of the presented system, Cortex can be linked as a library with the solver. Visualization of time–dependent data is not provided.

An interesting work in the field of time–dependent data visualization is UFAT [8]. Although this is only a particle tracer for unsteady flow fields, it is a good example to show processing of time–dependent data. It handles single and multi-block curvilinear grids, and the grid may have rigid-body motion. The data is read from files and processed sequentially, so no integrated environment is provided.

5 Conclusions and Outlook

A visualization system for arbitrary unstructured grids was presented. Solutions from finite volume (cell–centered) and finite element methods can both be visualized. They may be only piece–wise continuous. No other visualization system is able to handle this kind of data.

Using the heterogenous simulation data interface an easy connection of a numerical simulation and the visualization system can be made. This allows the integration of the visualization and a simulation application. Hence on–line simulation of the computed solution is possible. Also mechanisms for interactively steering the simulation are suggested. The output of an example session is presented.

Pre–processing can be integrated into this system. Data conversion modules from design systems to the HSDI–format will be developed. Then data conversion from the HSDI–format to the data structures used in the simulation must be provided to fulfill this task. This will lead to an integrated system, i.e. data flow from CAD system to CFD simulation and visualization and back.

On a further step the visualizer will be separated completely from the renderer. Therefore the interface between them must be extended. Then the visualizer can be parallelized using the DDD–library presented in [3]. This library allows to implement parallel applications, which are efficient and portable. Also the changes of the sequential code are minimal and done within a few days.

References

1. D. Banerjee, C. Morley, and W. Smith. The design and implementation of the cortex visualization system. In *Proccedings of the Visualization '94*, pages 265–272, Washington, D.C., October 1994. IEEE Computer Society Technical Committee on Computer Graphics in cooperation with ACM/SIGGRAPH, IEEE Computer Society Press.
2. D. Banerjee, T. Tysinger, and W. Smith. A scalable high-performance environment for fluid flows on unstructured grids. In *Proceedings of the Supercomputing '94*, pages 8–17, Washington, D.C., November 1994. IEEE Computer Society and ACM, IEEE Computer Society Press.
3. K. Birken. An efficient programming model for parallel and adaptive CFD-algorithms. In *Proceedings of Parallel CFD Conference 1994*, Kyoto, Japan, 1995. Elsevier Science.
4. A.-M. Duclos and M. Grave. Reference models and formal specification for scientific visualization. In Patrizia Palamidese, editor, *Scientific Visualization - Advanced Software Techniques*, Workshop Series, chapter 1.1, pages 3–14. Ellis Horwood, 1993.
5. M. Geiben and M. Rumpf. Visualization of finite elements and tools for numerical analysis. In *Proccedings of the Second Eurographics Workshop on Visualization in Scientific Computing*, Delft, Netherlands, April 1991.

34

6. Silicon Graphics, editor. *IRIS Explorer*^{TM} *2.0 Module Writer's Guide*. Silicon Graphics, Mountain View, California, 1992.

7. C. Helf and U. Küster. A finite volume method with arbitrary polygonal control volumes and high order reconstruction for the Euler equations. In S. Wagner, E.H. Hirschel, J. Périaux, and R. Piva, editors, *Proceedings of the Second European Computational Fluid Dynamics Conference*, Stuttgart, Germany, 1994. Wiley & Sons.

8. D. Lane. UFAT – a particle tracer for time–dependent flow fields. In *Proccedings of the Visualization '94*, pages 257–264, Washington, D.C., October 1994. IEEE Computer Society Technical Committee on Computer Graphics in cooperation with ACM/SIGGRAPH, IEEE Computer Society Press.

9. K. Polthier and M. Rumpf. A concept for time-dependent processes. In *Workshop Papers of the Fifth Eurographics Workshop on Visualization in Scientific Computing*, Rostock, Germany, May/June 1994.

10. A. Ruprecht. Finite elements for the calculation of turbulent flows in three-dimensional complex geometries. In *Proceedings of the 3rd International Congress of Fluid Mechanics*, Cairo, Egypt, 1990.

11. R. Struijs, P. Vankeirsbilck, and H. Deconinck. An adaptive grid polygonal finite volume method for the compressible flow equations. Meeting Paper AIAA-89-1959-CP, American Institute of Aeronautics & Astronautics, 1989.

Editors' Note: see Appendix, p. 151 for coloured figure of this paper

On a Unified Visualization Approach for Data from Advanced Numerical Methods

Martin Rumpf, Alfred Schmidt, Kunibert G. Siebert *
Institut für Angewandte Mathematik
Universität Freiburg
Hermann–Herder–Straße 10
D-79104 Freiburg i. Br., Germany

Abstract

Recent numerical methods to solve partial differential equations in scientific computing are based on a variety of advanced kinds of domain discretizations and appropriate finite dimensional function spaces for the solutions. The scope of grids under consideration includes structured and unstructured, adaptive and hierarchical, conforming and nonconforming meshes. The function spaces might be of Lagrangian or Hermitian type with higher polynomial degree and possibly discontinuous over element boundaries. Unfortunately, the rendering tools in scientific visualization are mostly restricted to special data structures which differ substantially from the data formats used in the numerical application. This forces users to map and interpolate their data, which is time consuming, storage extensive, and accompanied with interpolation errors.

We present an interface between numerical methods on various types of grids and general visualization routines which overcomes most of these disadvantages. It is based on a procedural approach managing a collection of arbitrary elements and a set of functions describing each element type.

1 Introduction

An important subject in scientific visualization is the integration of the visualization routines into the numerical code. The gap between the user's numerical data formats and the structures usually used by visualization tools has been realized as one of the fundamental outstanding problems in scientific visualization [14, 26]. Most of the visualization software [6, 10, 15, 25, 27] currently in use works on prescribed data formats. User data has to be converted into such a

---

*EMAIL: mart, alfred, kunibert@mathematik.uni-freiburg.de

format, for instance the UCD format implemented in AVS [1] or the HDF format from NCSA [16].

On the other hand a rapidly growing variety of advanced numerical methods to solve partial differential equations in scientific computing is based on various kinds of domain discretizations and appropriate numerical function spaces for the solutions.

To motivate our approach let us first review some of the mesh types occurring in recent numerical methods. The basic types are structured or unstructured meshes (from Finite Difference or Finite Element / Volume methods) [12]. To achieve a better approximation of singularities, such as shock waves, vortices, cracks or special boundary conditions, locally adapted meshes consisting of a single or of mixed element types, e. g. simplicial, prismatic, rectangular or cuboidal structure are in use [3, 23]. Local adaption on meshes of one element type, for instance rectangular grids, leads to nonconforming meshes, where the neighbourhood of elements across an element face is not one–to–one [2]. To achieve a good approximation of curved boundary segments, meshes with parametric (curved) elements have to be introduced. The polynomial order of the parametrization may be globally constant or locally changing from element to element. Finally hierarchically structured meshes have turned out to be very suitable for efficient numerical solvers [8].

The discrete solution belongs to some function space which might be of Lagrangian or Hermitian type [12] with higher polynomial degree when h–p–methods are considered [2]. In Finite Volume calculus the solutions are often only continuous inside the elements and have jumps over the element faces. Another example of a space of discontinuous functions is the one spanned by a non conforming divergence free basis for an incompressible velocity field [19].

In [22] we propose an approach, which tries to avoid restrictions on the ansatz functions and on the element types. A mesh is defined as a procedurally linked list of nonintersecting elements. Each single element is parametrized over a convex polyhedron with an arbitrary number of vertices and faces, which might vary from element to element. Only one element description is given for each element type. This description mainly contains a characterization of the reference shape and a set of functions evaluating coordinate transformations and element–to–element adjacencies. There is no random access to a single element. Information about elements is provided by a procedurally linked list. No permanent mapping of an in general enormous amount of numerical data onto new data structures is necessary. The visualization tools directly work on the data structures the user is accustomed to from his numerical method. He only has to provide some access procedures and give a description of the element types. A complete description of this concept can be found in [22]. Here we will give a brief outline, discuss some aspects and consequences for the implementation of specific visualization tools and give a couple of examples.

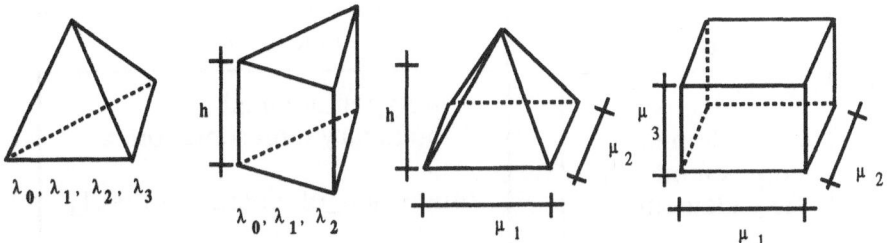

Fig. 1: Some possible local coordinate systems of a tetrahedron (barycentric coordinates $\{\lambda_i\}_{i=0,\cdots,3}$), a prism (barycentric coordinates $\{\lambda_i\}_{i=0,\cdots,2}$ on the base triangle and height h), a pyramid (bilinear coordinates μ_1, μ_2 on the base and height h) and a cube (trilinear coordinates μ_1, μ_2, μ_3)

2 The Concept's Starting Point

We are looking for an *unified interface* between numerical data of different types, at least those mentioned above, and general visualization tools. The interface should *not require the conversion to any new prescribed data formats* and *no additional storage* should be necessary. The interface should solely include *information* about the mesh *which is definitely needed* by the visualization routines. Finally, a user should be able to work solely on the data structures of his numerical methods.

Let us now analyze the type of access to the geometry and the function data that a general visualization method requires. We observe that there are primarily two types of element access. Elements may be processed in a *list–like order* when looping over all elements or following a path in the domain and crossing successively element faces to *pass* from one element *to one of its direct neighbours*. None of the usual methods needs a random access to an arbitrary element of the mesh.

Concerning the access to function data we realize that Finite Element / Finite Volume data is usually given by coefficients to basis functions naturally defined on the elements in *local coordinates*. A function f on a specific element e can easily be evaluated by a call like $f(e, c)$ where c is the local coordinate vector. In Fig. 1 we show some local coordinate systems for different types of elements. Furthermore, the mapping from global to these local coordinates and vice versa should be supported by specific *transformation routines*.

3 The Proposed Visualization Interface

We present only the structures for three dimensional meshes. The corresponding structures for two dimensional meshes are straightforward and quite similar; the main difference is that the description of 2D elements is a lot simpler. There are mainly three ingredients of this concept, the *Element3d* and *Element3d_Description* structures for a single element and the *Mesh3d* structure for the access

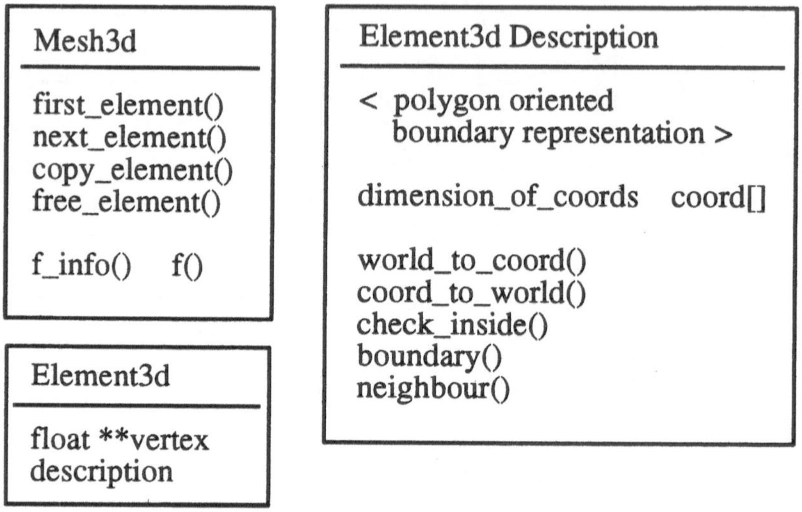

```
┌─────────────────────────────┐   ┌────────────────────────────────────────┐
│  Mesh3d                     │   │  Element3d Description                   │
│  ─────────                  │   │  ──────────────────                      │
│                             │   │  < polygon oriented                      │
│  first_element()            │   │     boundary representation >            │
│  next_element()             │   │                                          │
│  copy_element()             │   │  dimension_of_coords    coord[]          │
│  free_element()             │   │                                          │
│                             │   │  world_to_coord()                        │
│  f_info()    f()            │   │  coord_to_world()                        │
└─────────────────────────────┘   │  check_inside()                          │
                                   │  boundary()                              │
┌─────────────────────────────┐   │  neighbour()                             │
│  Element3d                  │   └────────────────────────────────────────┘
│  ─────────                  │
│  float **vertex             │
│  description                │
└─────────────────────────────┘
```

Fig. 2: The structures defining an arbitrary 3D mesh

to elements and function values on the geometry.

The geometry of a single element is described using a polygon–oriented boundary representation of the parameter domain. As we support only (curved) polyhedral elements, an element can be described by the set of its vertices and a description of the boundary polygons. This applies only to the local coordinates of an element; its shape in world coordinates is determined by some transformation routine. For a threedimensional polyhedron, we specify the number of boundary polygons (faces) and for each of these polygons the number of vertices, the local vertex indices and their order (thus giving an orientation to the polygon such that the surface normal is the outer normal to the polyhedron) and the local indices of the adjacent faces across the edges of each face. For each vertex, the coordinates in the local coordinate system of the element are given by *coord[]* in the *Element3d_Description*. Based on these values, the visualization routines will be able to operate in the local coordinate space. The functions *world_to_coord()* and *coord_to_world()* support the transformation between local coordinates and world coordinates. The pointer to a function *check_inside()* is provided which checks whether a point in local coordinates is inside the element or not. The element description is completed by pointers to routines *boundary()* and *neighbour()* which give information about the neighbourship of elements. The number of *Element3d_Description* structures equals the number of different element types.

As mentioned above, we use a procedural access to single elements in our mesh concept. The procedures return element data in a small structure *Element3d*. This structure for an element of the mesh mainly consists of a pointer to an *Element3D_Description*, which contains all information described above, and a vector *vertex* of pointers to the world coordinates of the element's vertices. Such information may be enlarged by optional global indexing information for the element itself and its vertices, which will be skipped here.

The whole mesh is a collection of elements together with an optional data function on the mesh. A function *first_element()* returns the first element and with *next_element()* we can successively walk through an arbitrary ordering of the elements. Both the *next_element()* and *neighbour()* routines may overwrite the current *Element3d* data structure, so that no additional storage is needed for the next or neighbouring element's data structures. In case that one needs to collect information about several elements, a routine *copy_element()* gives a copy of an element, which can be deleted later by a call of *free_element()*. This procedural access to the mesh elements allows the generation of local element data at the time when it is actually needed. Otherwise, in case of an array or pointered list, complete information for all elements of the mesh would have to be present at the same time (and occupy much more storage). The routines which generate the element data structures have to convert between the representations via user's data structures and the element structures. They do this mainly by refering to a previously prepared description and filling out vertex information for an element. In case of structured meshes this is done typically by index arithmetic, whereas for unstructured meshes such information is usually stored with the mesh.

Finally, the *Mesh3d* structure contains an interface to data given on the mesh. Here we want to support a situation where at the same time different types of data are given on the same mesh with different dimensions and other characteristics (example: a piecewise linear and scalar pressure, a piecewise quadratic and vector–valued velocity and some more), together with some method to select one of the data values for display. The data characteristics may vary from element to element in the same mesh, for example the polynomial degree of the ansatz functions. All these situations are managed by a rather general interface for functions. Instead of describing functions in terms of a Lagrangian basis, compare for instance [7], in our concept a function pointer *f(element, coord, value)*, supplied by the user, evaluates any type of user function on a specific *element* at a point in local coordinates *coord* and returns the possibly vector–valued function value in *value*. A function *f_info()* returns a structure filled with information about the function *f*, such as name, return value size and local polynomial order.

4   Evaluation of Visualization Requirements

In the last paragraph we propose structures to describe meshes and function data on them. This approach allows to work directly on the user data structures. We now have to reason whether the properties of this interface are sufficient to implement efficient visualization methods. Let us briefly pick up two fundamental display tools which are frequently used to visualize scientific data in 3D. We will thereby point out that the set of functions describing a single element and the procedures managing a mesh by giving access to its elements form a natural and minimal set of routines to deal with arbitrary meshes.

Calculating *isosurfaces* of a scalar quantity is one of the most popular visualization techniques. The straightforward approach is to loop over all elements in the mesh and check whether there is an intersection or not. Obviously this can be done using the procedures *first_element()* and *next_element()*. Taking into account the functions in the *Element_Description* structure handling adjacencies we realize that also more advanced approaches, for instance using extreme graphs as in [11], can be applied. Once having found an intersected element, we can reconstruct the local intersection using function evaluations in local coordinates and the description of the polygonal oriented boundary representation. Based on this data structure, several algorithms to solve the well known ambiguity problem can be implemented [17].

Visualizing *particle traces*, *streamlines*, *streamsurfaces* or moving *test volumes* is frequently used to emphasize the structure of flow phenomena [4, 13, 21]. Similar integration techniques are applied by strategies to extract topological information from simulation data sets [5, 9]. Based on some higher order ODE solver, a path in the computational domain in calculated. The right hand side of the ODE is any three dimensional field, velocity or eigenvectors, which has to be evaluated at points on or near the polygonal path. To keep the algorithm efficient we need a step by step tracking of the elements passed through. In more detail, to determine the neighbouring element of one specific face, to check the inclusion of a specific point in an element and to switch between global and local coordinates are the essential operations, which are requested. Exactly these are supported by the interface described above.

5   Results and Applications

This concept has been implemented in the programming environment GRAPE [28]. In comparison to dedicated software, a CPU overhead is produced by the general mesh concept. For our applications, this has turned out to be acceptable. We will now briefly sketch the application to data from different numerical methods.

Finite Element mesh moving in time: Here we have depicted two time steps from a (rotationally symmetric) simulation of the mean curvature flow of a torus, computed with the dynamic mesh algorithm from [18] based on an adaptive mesh generation. The numerical method only knows about the colored part of the mesh, which expands in front of the surface and is reduced behind it. The procedural approach we describe here is able to handle this configuration. On the right, isolines of the data in the transition zone and the elements at the boundary of this zone are drawn at two distinct times.

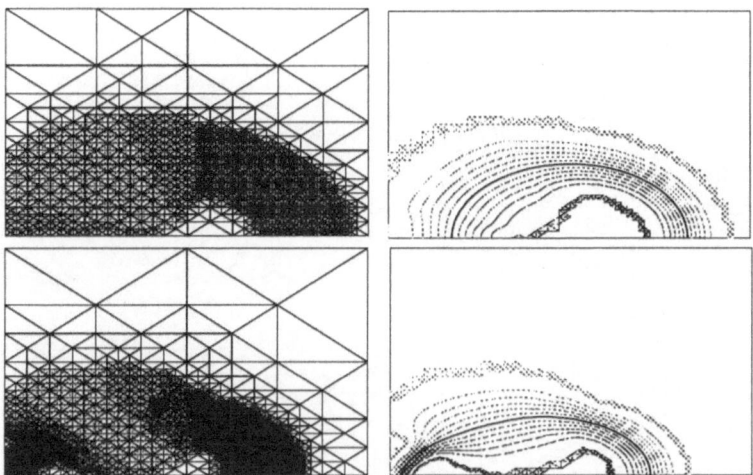

The moving surface is given by the zero level line On the left, we show the corresponding discretization of the whole domain solely handled by the independent mesh generator.

Finite Difference CFD computation: The vortex breakdown behind a cylindrical obstacle using a Finite Difference method is calculated. This computation was done at the Aerodynamisches Institut of Prof. Krause at the RWTH Aachen. We would like to thank for the opportunity to test our visualization tools on this data. The images show a coarse version of the underlying 113x41x33 Finite Difference grid, two slices with color shaded representations of energy and pressure and a pressure isosurface (see Appendix Fig. 3).

Finite Volume Mesh: The flow in a two stroke engine is a typical example for a problem with time dependent geometry [29]. One way to discretize the gas volume in the cylinder is to cut all elements of an initially tedrahedral mesh at the moving piston temporarily. Then "virtual" prismatic and tetrahedral elements appear in the boundary region (see Appendix Fig. 4). On the left a schematic 2D slice picks up some elements near the boundary and on the right using color shading the entropy is visualized on slices.

Prismatic and mixed type Finite Element Meshes: The third application is a Finite Element computation of the electric potential in a particle detector [24]. For the computation, a locally refined prismatic grid is used (see Appendix Fig. 5). Due to the presence of non–conforming nodes produced by the local refinement we may either build a conforming closure of the grid involving pyramids and tetrahedra [23] (on the left) or we may use a constrained approximation on the nonconforming purely prismatic grid (on the right with color shaded potential).

42

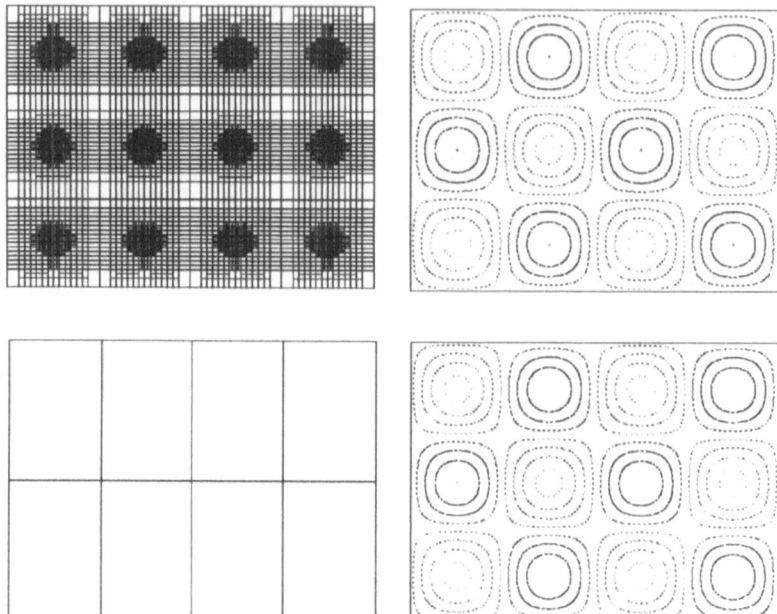

Finite Elements of h–p type: Finally we compare a nonconforming adaptive method (above) and an h–p–method (below), adaptive in space and with adaptive polynomial order up to 6, for the Poisson equation $-\Delta u = f$ with periodic f. On the left the two meshes and on the right isoline images of the corresponding solutions are drawn.

6   Future Work

In [20, 21] we propose a concept to deal with time dependent data, probably adaptive in time. We will combine these results with those discussed here. Our concept can also be applied in the context of distributed computing. The element access functions could address mesh elements on different machines or in a cluster of parallel processors. It would then no longer be necessary to hold the complete mesh information in the storage of the graphics workstation.

Furthermore, an important question is how to handle hierarchical data from multigrid methods. Switching between different levels is already possible with our interface in a user controlled manner. Taking into account the hierarchical structure and some information about bounds of different frequencies corresponding to different hierarchical levels, visualization methods can be tuned to a significant increase of the efficiency. We have started to consider these topics.

Acknowledgment

The development of the described algorithms and their realization in the software environment GRAPE would not have been possible without the support of the people at the SFB 256 graphics lab at Bonn University and at the Institut für Angewandte Mathematik at Freiburg University.

References

[1] Advanced Visual Systems, Inc.: AVS user's guide, Waltham, 1992

[2] Demkowicz, L., Oden, J. T., Rachowicz, W., Hardy, O.: Toward a universal h–p adaptive finite element strategy, Part 1 – Part 3. Comp. Meth. Appl. Mech. Engrg. 77, 79–212, 1989

[3] Bänsch, E.: Local mesh refinement in 2 and 3 dimensions, IMPACT Comput. Sci. Engrg. 3, 181–191, 1991

[4] Bryson, S.; Levit, C.: The Virtual Wind Tunnel, IEEE CG&A, 7, 25–34, 1992

[5] Delmarcelle, T.; Hesselink, L.: The Topology of Symmetric, Second–Order Tensor Fields, Proc. IEEE Visualization '94, 140–147

[6] Dyer, D. S.: A dataflow toolkit for visualization, IEEE CG&A 10, No. 4, 60–69, 1990

[7] Haber, R. B.; Lucas, B.; Collins, N.: A data model for scientific visualization with provisions for regular and irregular grids, Proc. IEEE Visualization '91

[8] Hackbusch, W. (ed.): Robust Mutli-Grid Methods, Notes on Numerical Fluid Mechanics, Vieweg, Braunschweig, 1988

[9] Helman, J. l.; Hesselink, L.: Visualizing Vector Field Topology in Fluid Flows, IEEE CG&A, 5, 36–46, 1991

[10] IBM, Inc.: IBM AIX Visualization Data Explorer, user's guide, IBM Publication SC38-0081

[11] Itoh, T.; Koyomada, K.: Isosurface Generation by Using Extreme Graphs, Proc. IEEE Visualization '94, 77–83

[12] Kardestuncer, K.: Finite element handbook, McGraw-Hill, New York, 1987

[13] Lane, D. A.: UFAT – A particle Tracer for Time-dependent Flow Fields, Proc. IEEE Visualization '94, 257–264, 1994

[14] Lang, U.; Lang, R.; Rühle, R.: Integration of visualization and scientific calculation in a software system, Proc. IEEE Visualization '91

[15] Lucas, B.; et. al. : An architecture for a scientific visualization system, Proc. IEEE Visualization '92

[16] NCSA HDF specification manual, available via anonymous ftp from ftp.ncsa.uiuc.edu

[17] Ning, P.; Bloomenthal, J.: An evaluation of Implicit Surface Tilers, IEEE CG&A, 11, 33-41, 1993

[18] Nochetto, R. H.; Paolini, M.; Verdi, C.: A dynamic mesh algorithm for curvature dependent evolving interfaces, J. Comput. Phys. (to appear)

[19] Pironneau, O.: Méthode des Éléments Finis pour les Fluides, Masson, Paris, 1988

[20] Polthier, K.; Rumpf, M.: A Concept for Timedependent Processes, Proc. Eurographics Workshop 94, Rostock, 1994

[21] Rumpf, M.; Geiben, M.: Moving and tracing in timedependent vector fields on adaptive meshes, Report, SFB 256, Bonn, 1994

[22] Rumpf, M.; Schmidt, A.; Siebert, K. G.: Functions defining arbitrary meshes, a flexible interface between numerical data and visualization routines, Report, SFB 256, Bonn, submitted to Computer Graphics Forum

[23] Siebert, K. G.: Local refinement of 3d–meshes consisting of prisms and conforming closure, IMPACT Comput. Sci. Engrg. 5, 271–284, 1993

[24] Siebert, K. G.: An a posteriori error estimator for anisotropic refinement, Preprint 313, SFB 256, Bonn, 1993 (to appear in Num. Math.)

[25] Silicon Graphics Computer Systems, Inc.: IRIS Explorer, Tech. Report BP–TR–1E–01, 1991

[26] Treinish, L. A.: Data structures and access software for scientific visualization, Computer Graphics 25, 104–118, 1991

[27] Upson, C.; et. al.: The Application Visualization System: A computational environment for scientific visualization, IEEE CG&A 9, No. 4, 30–42, 1989

[28] Wierse, A.; Rumpf, M.: GRAPE, Eine objektorientierte Visualisierungs– und Numerikplattform. Informatik Forschung und Entwicklung 7, 145–151, 1992

[29] Wierse, M.: Higher order upwind schemes on unstructured grids for the compressible euler equation in timedependent geometries in 3d. Dissertation, Freiburg, 1994

Editors' Note: see Appendix, p. 152 for coloured figures of this paper

Raycasting with Opaque Isosurfaces in Nonregularly Gridded CFD Data

Thomas Frühauf

Darmstadt Technical University
Department of Computer Science, Interactive Graphics Systems Group
Wilhelminenstrasse 7, 64283 Darmstadt, Germany
email: thfrueha@igd.fhg.de

Abstract: Direct volume rendering (DVR) is becoming more and more useful for the graphical analysis of computational fluid dynamics (CFD) data, because of the extremely huge datasets that are generated with today's supercomputers. However, only semi-transparent visualisations have been produced via DVR from nonregularly gridded simulation data so far. Such images provide no information about the data distribution in the viewing direction. This paper reports our realisation of opaque and combined semi-transparent/opaque raycasting in nonregular grids. We emphasize on the mapping process, on the colour accumulation, and on the shading of isosurfaces. The generated images provide both holistic information and cues about the spatial data distribution in the viewing direction. We use the same interpolating functions in the visualisation algorithms as they are used in the data generation with the Finite Element method. Therefore, the rendered isosurfaces reveal interesting features that cannot be seen when isosurfaces are extracted directly with simplier algorithms.

Keywords: *Direct Volume Rendering, Raycasting, Opaque Rendering, Isosurfaces, Nonregular Grids, Finite Element Data, Computational Fluid Dynamics*

1 Motivation

Steady advances of simulation software as well as new generations of numerical processors allow for more and more complex numerical studies of real-life problems, such as weather forecasts, the flow around airplanes, or the distribution of industrial polutants. Such computational fluid dynamics (CFD) data is usually organized on huge curvilinear or unstructured numerical grids [1]. Nonregular numerical grids help solving huge simulation tasks with less computational effort than regular grids can, because less elements are required to cover complex shaped volumes. However, nonregular grid elements require more computational efford in the analysis phase, i.e., the visualisation.

The interpolation of nonregularly gridded volume data onto a regular grid enables the use of fast visualisation algorithms (at first sight). It destroys, however, existing context information and has negative effects on the size and the quality of the data [2]. Therefore, efficient visualisation algorithms for huge nonregularly gridded volume data are needed [3]. Among the possible techniques are 'isosurface extraction (ISE)' and 'direct volume rendering (DVR)'.

In ISE, a surface of constant value is extracted and displayed using traditional geo-metric rendering. A sequence of isosurfaces, from minimum to maximum value can be animated in order to gain an impression of the entire volume. Although a single shaded isosurface can be *rendered* in realtime on modern workstations, the *computation* of isosurfaces in huge nonregular volumes (> 200,000 Finite Elements) cannot be done in real-time [4]. Therefrom, the mental reconstruction of the whole volumetric data distri-bution is hindered in interactive visualisation systems. Besides, the faster methods, such as "Marching Cubes" [5] do not represent the data very accurately (as described below) when they are applied to Finite Element Data.

Via DVR the whole data can be displayed in a holistic manner in one single image. But the images are often fuzzy and do not provide any depth cues. Raycasting [e.g., 6, 7, 8, 9], allows the opaque rendering of virtual surfaces in volume data, as it is often utilized in medical imaging[1]. With a suitable mapping transfer function, both semi-transparent and opaque regions can be displayed together in one image. However, the literature on the raycasting of nonregular grids is still sparse [14, 15, 16, 17, 18], and - to the author's knowledge - there is no work published yet on the opaque DVR of data on nonregular grids. The aim of the work described herein was to realize combined semi-transparent/opaque raycasting on nonregular grids. The driving intention was, to generate more accurate visualisations of numerical simulation data with a DVR tech-nique. Therefore, we used the same interpolating functions in the visualisation process as they are used in the data generation with the Finite Element method.

The remainder of the paper is structured as follows: In the next section we recall two strategies for the fast tracing of rays through nonregular grids as well as the methods for local data sampling there. In chapter three we introduce our mapping stra-tegy, the applied material model, and the algorithm for the accumulation of the pixel colour of the final image. Chapter four describes how opaque isosurfaces are visuali-zed with our raycasting implementation and how we estimate local surface normals in nonregular grids for the shading of the isosurfaces. Finally, in chapter five, we discuss the visual appearence of isosurfaces visualized with our method.

2 Baseline: Raytracing and Interpolation in Nonregular Grids

Raycasting in nonregular grids follows the same strategy like raycasting of regular voxel grids. However, finding the rays' traces as well as sampling along the rays is more computing-intensive due to the complex grid topology. In particular, the main problems are:

- The identification of the cell through which a ray enters the volume.

- The identification of the cell in which the next sampling point is located.

---

1. Projection methods [e.g. 10, 11, 12] allow, at least in theory, opaque virtual surfaces, too. But the surfaces coincide here - like in ISE by the cuberille method [13] - with cell faces, which gives isosurfaces a poor quality in the case of nonregular grids.

- The interpolation of the data to be visualized from the vertices of a non-regular cell.

The location of the cell face which is first intersected by a certain ray can be computed by testing the ray against the planes of all exterior cell faces. The right cell is the one, whose plane-intersection point has the smallest distance from the virtual starting point of the ray. This quite inefficient strategy can be enhanced by the embedding of the nonregular grid in a coarse regular grid: For each cell of the regular grid those cell faces of the nonregular grid are registrated which are, at least partially, located in this cube. This restricts the number of cell faces to be tested since a ray's trace through the regular grid is quickly found by a 3D Bresenham algorithm.

A much faster solution can be realized with the so-called 'item-buffering' technique [19]. Here, we 'colour-code' the exterior cell faces and render the grid conventionally, i.e., by hardware using the z-buffer. The mapping of cell numbers to colour is simple, e.g., in the RGB colour model, and can be easily inverted. Therefore, by scanning the pixels of the grid's image, we can quickly determine the cell faces through which the rays enter the nonregular grid.

Following the rays' traces through the cells is the most expensive part in the raycasting of nonregular grids. Firstly, we have to find that cell in which a subsequent sampling point is located; secondly, we have to perform a local interpolation from the data that is defined at the vertices of the nonregular cell primitives. Conventionally, these tasks are very computing-intensive and require the repeated transformation into the local coordinates of the cells along each ray. However, it is possible to speed-up this processes with dedicated solutions for unstructured grids of tetrahedra [14] as well as for curvilinear grids consisting of nonregular hexahedra [18].

Garrity proposes to pre-compute a cell adjacency list of an *unstructured grid* in order to follow the rays through the volume[1]. Data sampling should be done at the ray/cell intersection points by interpolation from the vertices of the particular cell face.

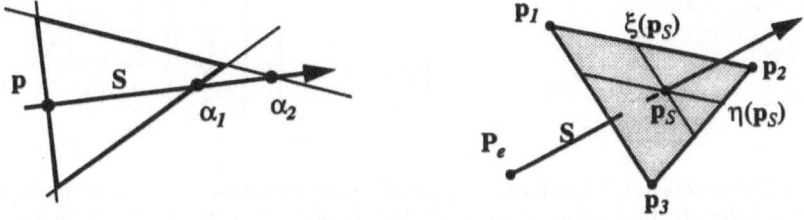

Fig. 1: Left: The exit face is determined by intersecting the ray with all faces' planes. (The principle is shown in a 2D sketch). Right: The determination of local coordinates $\xi(p_S)$, $\eta(p_S)$ in the exit face of a tetrahedron.

---

1. An algorithm that is linear to the number of cells is decribed in [20].

48

In order to determine that cell face through which a ray leaves a cell, the distance α of the intersection point with the face's plane from the enterance point P_e is computed for each cell face. The 'real' exit face is the one with the smallest positive distance a. Garrity gives the closed formulas for the computation of α as well as for the local coordinates ξ, η of the intersection point inside the exit face. It is then possible to perform a parametrical interpolation in the exit face. The data value $v(p_s)$ at the exit point is computed from the data $v(p_i)$ at the faces nodes with the *shape functions* of a linear interpolating triangle.

In the case of *curvilinear grids* which consist of nonregular hexahedra, it is not possible to compute the local coordinates in the exit face directly - a computing-intensive iterating algorithm would be nessesary for this task. This is due to the fact that a 8-node hexahedron has shape functions that are not linear in all their parameters. However, it is possible to transform the rays locally at any point of a curvilinear grid into the regular computational space (C-space) of a curvilinear grid (see figure 2).

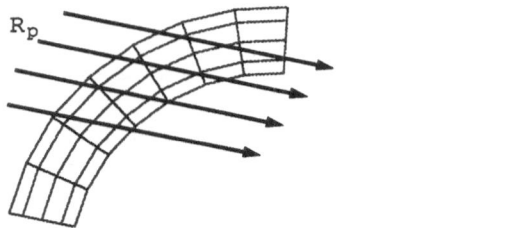 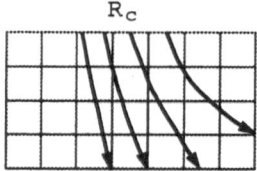

Fig. 2: The transformation of rays from physical space to computational space simplifies cell identification and interpolation along the rays.

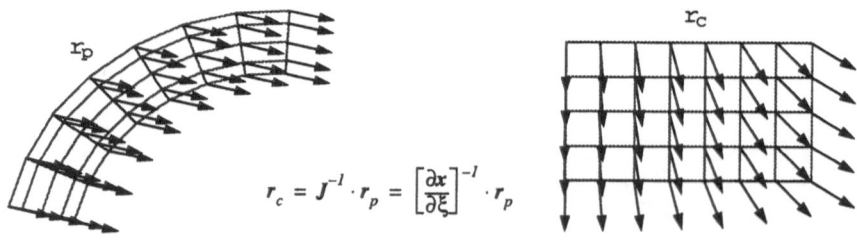

$$r_c = J^{-1} \cdot r_p = \left[\frac{\partial x}{\partial \xi}\right]^{-1} \cdot r_p$$

Fig. 3: The vectors r_p are transformed into the C-space. Direction and length of the r_c 's is determined by the local Jacobian and reflect thus the grid's local shape and warp.

We realize this operation by the pre-computation of a C-space vector r_c for each grid node from the rays in physical space r_p and the local *Jacobian J* (see figure 3). The Jacobian at inner grid nodes might be approximated with central differences. Using forward/backward differences [21] instead, enhances the accuracy during the integration of the rays' traces at the cost of storage space (we have to compute and store one vector r_c for each grid node of each element, i.e. eight times for an inner grid node). The tangential curves, also called the *streamlines*, of the vector field r_c coincide with the rays' traces in C-space. Thus, we can follow the rays by piecewise integration in this vector field (we achieved the best compromise between computational speed and accuracy with *Heun's scheme* (i.e., 2nd order Runge-Kutta). Data sampling is performed after each integration step via *trilinenar interpolation* in the regular cell of the C-space.

3 Image Generation in the Raycasting of Nonregular Grids

So far, we have just described the strategies to compute the rays' pathes through a non-regular grid and to interpolate along the rays for data sampling. However, the image generation by raycasting requires also the mapping from data to colour and opacity, a material model, and an algorithm for the accumulation of the final pixel colour.

3.1 Mapping

Raycasting demands for the specification of mapping transfer functions from data to colour and opacity. Instead of the RGB model we use here the more intuitive HLS (Hue, Lightness, Saturation) model [22]. We implemented a graphical user interface that allows the interactive definition, modification (and storage) of each transfer function through linear and cubic splines. We usually work with a constant function for *lightness* as well as for *saturation*. *Hue* values are adjusted with the aid of a data histogram individually for each dataset. Mostly, we use constant or 'staircase' functions for the opacity. Namely, we specify an isosurface through a C^{-1} continuous opacity transfer function (see section 4). The transfer functions are converted into four Look-Up Tables, each with 256 members. Therefore, we do not have to evaluate the spline function during the raycasting.

3.2 Material Model

In the mapping process of DVR, data values are converted to local colour and opacity. It has to be defined how these local values are interpreted in the accumulation of pixel colours. In particular, the model has to integrate the effect of the non-equal sampling distances as they result from the raytracing strategies described in section two.

In semi-transparent rendering we use the *homogeneous material model* that Wilhelms and van Gelder formulated for direct volume rendering through projection [23]. The model defines that "each cell emits its own light, transmits some light coming from behind, and occludes some light coming from behind and from within the cell". In this model *opacity O* and *lightness L* along a ray behave like:

$$O(s) = 1 - e^{-\Omega(s) \cdot s}, \qquad (1)$$

$$L(s) = \frac{\lambda(s)}{\Omega(s)} \cdot O(s). \qquad (2)$$

Here, s is the parameter along the ray. The *specific opacity* Ω of the medium and the *specific lightness* λ of the light that is emitted by the medium are defined through the mapping transfer functions[1]. Following a ray through the grid, we have to compute colour and opacity between two subsequent sampling points, i.e., we have to integrate along a distance $d = |p_2 - p_1|$. We assume a homogeneous medium between the sampling for which we average colour and opacity between the two points. Hence, we can compute the segment's opacity O_d and lightness L_d by:

$$O_d = 1 - e^{-\Omega_a \cdot d} \tag{3}$$

$$L_d = \frac{\lambda_a}{\Omega_a} \cdot O_d. \tag{4}$$

with $\Omega_a = 1/2 \cdot (\Omega(p_1) + \Omega(p_2))$ and $\lambda_a = 1/2 \cdot (\lambda(p_1) + I(p_2))$.

A speed-up can be achieved when we approximate $(1 - e^{-\Omega s})$ by $min(1, \Omega s)$. This results in:

$$O_d = min(1, (\Omega_a \cdot d)) \tag{5}$$

and for $\Omega_a \cdot d < 1$:

$$L_d = \lambda_a \cdot d. \tag{6}$$

Equations 5 and 6 tend to overvalue lightness and opacity compared to equ. 3 and 4. With real-life datasets we could, however, not detect great differences in the information content between the different visualisations. Therefore, most often we use the faster linearisation.

3.3 Accumulation of Pixel Colour

We have already included the effect of non-equal sampling distances in the calculation of the segments' colour and opacity. Therefore, the pixel colour C_P can be accumulated via alpha-blending from the k segments with segment colour c and segment transparency a:

$$C_P = \sum_{i=0}^{k} c_i (1 - \alpha_i) \prod_{j=0}^{i-1} \alpha_j \tag{7}$$

---

1. For reasons of intuitivity we specify in the Mapping-GUI an 'object opacity' $O_{Obj}[0, 1]$ which is converted to $\Omega[0, \infty]$ by: $\Omega = 1/\sqrt{n_e} \cdot log(1/(1 - O_{Obj}))$, with n_e as the number of grid cells.

Using opacities $O = 1 - \alpha$ - as in our implementation - instead of transparencies, we can derive the following algorithm for the accumulation of pixel colour:

$$C_{i+1} = (1 - O_i) \cdot C_d + C_i \qquad (8)$$

$$O_{i+1} = (1 - O_i) \cdot O_d + O_i. \qquad (9)$$

Here O_0 equals *zero*. We stop the acculumation when $(1 - O_{i+1}) < \varepsilon$, where ε is a small positive constant; then, any segments further down a ray would not alter the accumulated pixel colour substantially. If a ray leaves the grid, we finally add the backgroud colour $C_d = C_B$ with its background opacity $O_d = O_B = 1$.

Usually, we use one ray per pixel. It is, however, possible to scale-up previews which have been generated with a small number of rays. So far, we have not experimented with non-constant ray densities in one image. But we will investigate how this approach can further speed-up our raycasting implementation. E.g., we could choose a certain ray density according to registrated colour gradients between pixels of a preview.

4 Opaque Isosurfaces

The strategy to accumulate the pixel colours, as it was explained in the previous section, allows us to generate holistic views of the whole volume. The images, however, do not provide a pictorial cue for the spatial distribution of data in the viewing direction. This is due to the integrating characteristic of this approach (see sections 3.2 and 3.3). In contrast to the applied model of a semi-transparent and light-emitting material, opaque isosurfaces can be shaded by standard methods like *Gouraud* or *Phong shading*. The opaque isosurfaces, thus, provide the spatial information in the viewing direction. On the other hand, single isosurfaces show only one level in the data range. Our idea is to mix the semi-transparent and the opaque parts to provide both the holistic information about the whole data range and the spatial distribution of interesting data levels.

One approach for this task is to perform two independent rendering processes - one for the semi-transparent volume data and one for a polygonal isosurface - and to mix the results of both processes as described in [24]. Since our opaque objects - the isosurfaces - are related directly to the volume data, we follow a different strategy here. As mentioned before, we can use the raycasting algorithms described herein to render the opaque isosurfaces, too. We are aware of the fact that the mixing strategy is faster when we produce several images with isosurfaces at different data levels. On the other hand, we are able to extract the isosurfaces with a better quality (as explained below) by raycasting than by standard methods like "Marching Cubes"!

In order to generate the *combined semi-transparent/opaque visualisations*, we apply the homogeneous material model and perform a shading calculation for a ray

52

segment colour when an isosurface level - that has been defined with the mapping transfer functions - is crossed. The altered segment colour is then mixed with the accumulated pixel colour in the standard way (equ. 8, 9). Both the Gouraud and the Phong shading calculation demand for a surface normal. In the raycasting of voxel grids the direction of the function gradient $\nabla\Phi$ is often used for the normal of level-surfaces [25]. We use the function gradient here, too; in the following secting we explain how we determine the local function gradient quickly during raycasting in curvilinear and unstructured grids.

4.1 Estimation of Isosurface Normals in Nonregular Grids

The gradient $\nabla\Phi$ of a C^1 continuous function Φ is perpendicular to any isosurface of this function. In case of Finite Element simulations, Φ is C^1 continuous inside the elements, but only C^0 continuous across element boundaries. Thus, $\nabla\Phi$ is C^{-1} continuous across element boundaries. In the case of curvilinear grids, we pre-compute and store one gradient for each node of each element, i.e. eight gradients for an inner node[1]. During raycasting we interpolate these nodal gradients at the local coordinates of the sampling points by trilinear interpolation for a fast local gradient estimation (and *Phong shading*). In a tetrahedron an isosurface is planar and the surface normal is constant. Consequently, here, the gradient is stored for each cell and no interpolation is performed at the sampling point.

The function Φ inside a finite element is defined from the function values f_i at the element nodes and the shape functions N_i by:

$$\Phi(x) = \sum_{i=1}^{n} N_i(\xi(x)) \cdot f_i. \qquad (10)$$

We want to calculate the function gradient:

$$\nabla\Phi = \left[\frac{\partial\Phi}{\partial x}\ \frac{\partial\Phi}{\partial y}\ \frac{\partial\Phi}{\partial z}\right]^T. \qquad (11)$$

Applying the Chain Rule, the derivative, say, with respect to x is:

$$\frac{\partial\Phi}{\partial x} = \sum_{i=1}^{n}\left(\frac{\partial N_i}{\partial\xi}\cdot\frac{\partial\xi}{\partial x}+\frac{\partial N_i}{\partial\eta}\cdot\frac{\partial\eta}{\partial x}+\frac{\partial N_i}{\partial\zeta}\cdot\frac{\partial\zeta}{\partial x}\right)\cdot f_i. \qquad (12)$$

Therefore, we have to compute the derivations of the shape functions with respect to

---

1. The alternative - if storage space was not available - would be to compute just one gradient for each grid node which would result in a C^0 continuous isosurface normal across cell boundaries.

the local coordinates. The partial derivatives of the local coordinates with respect to the global coordinates can be obtained from the inversion of the Jacobian J.

Defining a 'C-space gradient'

$$\nabla_c \Phi = \left[\frac{\partial \Phi}{\partial \xi} \; \frac{\partial \Phi}{\partial \eta} \; \frac{\partial \Phi}{\partial \zeta}\right]^T \tag{13}$$

we can write:

$$\nabla \Phi = J^{-1^T}(x) \cdot \nabla_c \Phi. \tag{14}$$

In curvilinear grids we compute J and $\nabla_c \Phi$ at an inner grid node i, j, k via combined forward/backward differences [21]. One of the eight cases, say, forward/backward/forward, is computed by:

$$J_{i,j,k} = \left[\frac{\partial x}{\partial \xi}\right] \approx \left[\frac{\Delta x}{\Delta \xi}\right] = \begin{bmatrix} x_{i+1,j,k} - x_{i,j,k} \\ x_{i,j,k} - x_{i,j-1,k} \\ x_{i,j,k+1} - x_{i,j,k} \end{bmatrix}^T \tag{15}$$

$$\nabla_c \Phi_{i,j,k} = \left[\frac{\partial \Phi}{\partial \xi}\right] \approx \left[\frac{\Delta \Phi}{\Delta \xi}\right] = \begin{bmatrix} f_{i+1,j,k} - f_{i,j,k} \\ f_{i,j,k} - f_{i,j-1,k} \\ f_{i,j,k+1} - f_{i,j,k} \end{bmatrix} \tag{16}$$

In unstructured, tetrahedronal grids the gradient is constant over the whole area of a tetrahedron (see equ. 18, 19), since the shape functions N_i of a 4-node tetrahedron are linear. Thus, an isosurface is planar inside a tetrahedron:

$$\begin{aligned} N_1 &= \xi \\ N_2 &= \eta \\ N_3 &= \zeta \\ N_4 &= 1 - \xi - \eta - \zeta \end{aligned} \tag{17}$$

54

Therefore, we get:

$$J = \left[\frac{\partial x}{\partial \xi}\right] = \sum_{i=1}^{n} \frac{\partial N_i}{\partial \xi} \cdot x_i = \begin{bmatrix} x_1 - x_4 \\ x_2 - x_4 \\ x_3 - x_4 \end{bmatrix}^T \qquad (18)$$

$$\nabla_c \Phi = \left[\frac{\partial \Phi}{\partial \xi}\right] = \sum_{i=1}^{n} \frac{\partial N_i}{\partial \xi} \cdot f_i = \begin{bmatrix} f_1 - f_4 \\ f_2 - f_4 \\ f_3 - f_4 \end{bmatrix} \qquad (19)$$

4.2 'Exact' Detection of Isosurfaces

During the sampling along the rays we usually do not hit an isosurface exactly at a certain sampling point. In most cases we find a function value below/above the critical value Φ_S at a sampling point x_{i-1}, while Φ is above/below that value at the next sampling point. For a better interpolation of the sample colour and the isosurface's normal vector we step back along the ray to a position x_s. We assume a linear variation of Φ along the ray for the determination of that position[1]:

$$x_S = x_{i-1} + \left(\frac{\Phi_S - \Phi_{i-1}}{\Phi_i - \Phi_{i-1}} \cdot (x_i - x_{i-1}) \right). \qquad (20)$$

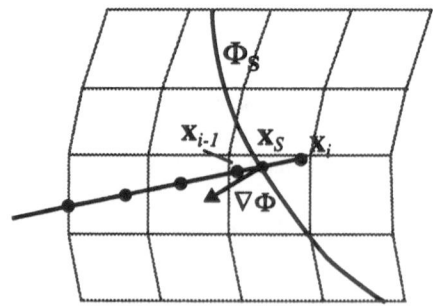

Fig. 4: The determination of the position x_s on the isosurface from the function values at two sub-sequent sampling positions

In the piecewise sampling along the rays it is theoretically possible to miss an isosurface if it encloses a very thin volume. Hence, we reduce our sampling distances if at

---

1. We are aware of the fact that this is a simplification in the case of curvilinear grids.

least one of an element's nodal data is above and one is below the isosurface level, which indicates that an isosurface is existent in this element.

We found it useful to store the detected isosurface normals together with the accumulated pixel colours of the final image. This allows the interactive variation of the directed light source after the raycasting algorithm. For a new shading calculation we just have to execute a loop over all pixels. Therefore, we can interactively find the best light position to reveal the 3D shape of an isosurface in a still image.

5 Discussion

Opaque raycasting of curvilinearly gridded data reveals features that are not visible when isosurfaces are extracted with an implementation of "marching cubes (MC)" for curvilinear grids (see color plate 1). Furthermore, the combination of semi-transparent and opaque rendering provides additional information. The different shapes of the isosurfaces can be explained when we investigate the underlying interpolating functions. In MC, data is interpolated linearly along cell edges, and planar surface patches are then constructed due to Look-Up Tables. In standard implementations, as they are realized in many visualisation systems (also in well-known commercial systems), no further inspection of data inside the cells is done. Thus, holes or 'wrong' connections are possible which have been discussed in many papers following the original MC publication. Therefore, a feature as it is enlarged in colour plate 2 is easily explained.

In our view the concave and convex segments of the raycasted isosurface are more interesting. They reflect the cubical interpolating functions and the C^0 continuity of data at element faces in Finite Element grids. Since we have used the same interpolating functions for the surface extraction and surface shading, these features are also visible in our images.

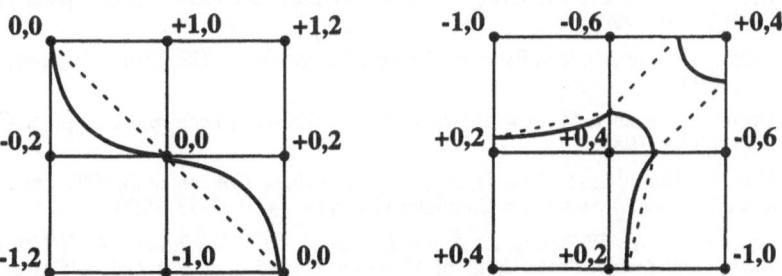

Fig. 5: Possible isosurface shapes (level 0.0) in Finite Element data. Characteristic features can be missed when using simple isosurface extraction techniques (dashed lines).

We are aware of the fact that specific visualisations are required for individual addressees. Therefore, it can even be necessary to graphically smooth partially planar isosurfaces at segment edges if a broader public is addressed. On the other hand, if the

56

interpolating functions of the data generating process are known (like in CFD data) and the results are to be inspected by the researchers, we should use the same interpolating functions in the visualisation algorithms.

The purpose of visualisation is insight, not pretty pictures[1].

Acknowledgements

I would like to thank J. van Wijk for early comments on the calculation of surface normals. Furthermore, I am greatful to P.G. Bunning for writing his inspiring paper „Sources of Error in the Graphical Analysis of CFD results" [27]. Last but not least I thank H. Rettig for programming work and S. Wurster for proof reading this paper.

References

1. Speray, D.; Kennon, S.: *Volume Probes - Interactive Data Exploration on Arbitrary Grids.* Computer Graphics, 24(5):5-12, 1990.

2. Wilhelms, J.; Challinger, J.; Alper, N.; Ramamoorthy, S.; Vaziri, A.: *Direct Volume Rendering of Curvilinear Volumes.* Computer Graphics 24(5):41-47, 1990.

3. Kaufman, A.; Höhne, K.; Krüger, W.; Rosenblum, L.; Schröder, P.: *Research Issues in Volume Visualization.* IEEE Computer Graphics & Applications, 14(2):63-67, 1994.

4. Ellsiepen, P.: *Parallel Isosurfacing in Large Unstructured Datasets.* In: Göbel, M.; Müller, H.; Urban, B. (eds.): *Visualisation in Scientific Computing.* pp. 9-23, Springer Verlag, Wien, 1995.

5. Lorenson, W.; Cline, H.: *Marching Cubes: A high resulution 3D surface construction algorithm.* Computer Graphics, 21(4):163-169, 1987.

6. Kajia, J.; von Herzen, B.: *Ray-tracing volume densities.* ACM Computer Graphics, SIGGRAPH 84, 18:165-174, 1984.

7. Levoy, M.: *Display of Surfaces From Volume Data.* IEEE Computer Graphics & Applica-tions, 8(3):29-37, 1988.

8. Krüger, W.: *Volume Rendering and Data Feature Enhancement.* Computer Graphics, 24(5):21-26, 1990.

9. Sakas, G.: *Interactive volume rendering of large fields.* The Visual Computer, 9:425-438, 1994.

10. Upson, G.; Keeler, M.: *The V-Buffer: Visible Volume Rendering.* Computer Graphics, 22(4):59-64, 1988.

11. Max, N.; Hanrahan, P.; Crawis, R.: *Area and Volume Coherence for Efficient Visualisation of 3D Scalar Functions.* Computer Graphics, 24(5):27-33, 1990.

12. Van Gelder, A.; Wilhelms, J.: *Rapid Exploration of Curvilinear Grids Using Direct Volume Rendering.* Proceedings of Visualization '93, San Jose Cal., USA, pp. 70-77, 1993.

13. Chen, L.: *Surface Shading in the Cuberille Environment.* IEEE Computer Graphics and Applications, Dec. 1985, pp. 33-43, 1985.

14. Garrity, M.: *Raytracing Irregular Volume Data.* Computer Graphics, 24(5):35-40, 1990.

---

1. After R. Hamming: „The purpose of computing is insight, not numbers" [26]

15. Ramamoorthy, S.; Wilhelms, J.: *An Analysis of Approaches to Ray-Tracing Curvilinear Grids*. University of California at Santa Cruz, Technical Report UCSC-CRL-92-07, 1992.

16. Koyamada, K.; Nishio, T.: *Volume Visualisation of 3D Finite Element Method Results*. IBM Journal of Research and Development, 35(1/2):12-25, 1991.

17. Tabatabai, B.; Sessarego, E.; Mayer, H.: *Volume Rendering on Non-regular Grids*. Computer Graphics Forum 13(3):247-258 (Eurographics 94), 1994.

18. Frühauf, T.: *Raycasting of Nonregularly Structured Volume Data*. Computer Graphics Forum 13(3):295-303 (Eurographics'94), 1994.

19. Weghorst, H.; Hooper, G.; Greenberg, D.: *Improved Computational Models for Ray Tracing*. Transaction on Graphics; 3(1):52-69, 1984.

20. Frühauf, T.: *Interactive Visualisation of Vector Data in Unstructured Volumes*. Computers & Graphics, 18(1), 1994.

21. Sadarjoen, A.; van Walsum, T.; Hin, A.; Post, F.: *Particle Tracing Algorithms for 3D Curvilinear Grids*. In Proceedings: 5th Eurographics Workshop on Visualization in Scientific Computing, 30.5.-1.6. 1994, Rostock, 1994.

22. Foley, J.; van Dam, A.; Feiner, S.; Hughes, J.: *Computer Graphics: Principles and Practice (2nd edition)*. Adison Wesley, Reading, Mass., 1990.

23. Wilhelms, J.; Van Gelder, A.: *A Coherent Projection Approach for Direct Volume Rendering*. Computer Graphics (SIGGRAPH'91), 25(4):275-284, 1991.

24. Frühauf, M.: *Combining Volume Rendering with Line and Surface Rendering*. In: Post, F.; Barth, W. (eds.): *Eurographics'91*, pp. 21-32; North Holland, Amsterdam, 1991.

25. Höhne, K.; Bernstein, R.: *Shading 3D-images from CT Using Gray-level Gradients*. IEEE Transaction on medical imaging, MI-5(1):45-47, 1986.

26. Hamming, R.: *Numerical Methods for Scientists and Engineers*, Mc Graw-Hill, New York, 1962.

27. Buning, P.: *Sources of Error in the Graphical Analysis of CFD Results*. Journal of Scientific Computing, 3(2), 1988.

Editors' Note: see Appendix, p. 153 for coloured figures of this paper

On the Optimization of Projective Volume Rendering

P. Cignoni[†], C. Montani[‡], D. Sarti[‡], R. Scopigno[*]

[†] Dip. Scienze dell'Informazione – Univ. Pisa, C.so Italia, 40 - 56100 Pisa, Italy
[‡] I.E.I. – C.N.R., Via S. Maria 46, - 56126 Pisa, Italy
[*] CNUCE – C.N.R., Via S. Maria 36, - 56126 Pisa, Italy

Abstract. How to render very complex datasets, and yet maintain interactive response times, is a hot topic in volume rendering. In this paper we focus on projective visualization of datasets represented via tetrahedral tessellations. Direct projective visualization is performed by sorting tetrahedra with respect to view direction and then by projecting them onto the screen. Different sorting algorithms and "per tetrahedra" projection techniques are reviewed and evaluated. A new method for tetrahedra projection approximation is presented. In addition, we compare the results obtained by the optimization of the rendering process with those obtained by adopting a data simplification approach.

1   Introduction

The real usability of a system for the visualization of volume data is strictly connected to the level of interactivity the system performs. This is both to simplify the user–system interaction and to improve our understanding of the results through motion or animation. The efficiency of the visualization algorithm is therefore crucial.

Direct volume rendering techniques [3] are generally characterized by the high complexity of the rendering process. Volume datasets can be visualized by applying either projective or ray casting methods. In the case of curvilinear or unstructured datasets, a projective approach using hardware rendering features (polygon scan conversion, shading and compositing) is much more efficient. A comparison of projection *vs.* ray casting gave much shorter times for the first approach (from 5 to 100 times faster, on the same dataset, depending on the optimization techniques applied [2]). Also, the images produced with a projective approach are very close in quality to those generated by a ray casting algorithm.

This paper thus analyzes and evaluates several optimization techniques to speed up projective visualization. We consider algorithms which work on datasets represented using tetrahedral decompositions. Tetrahedral decomposition is a suitable unifying representation, at least for curvilinear or unstructured volume datasets, because:

- visualization software, based on simplicial elements, is very general: it can also be used to render curvilinear or regular datasets (the decomposition of hexahedral cells into tetrahedra is easy). Such a decomposition increases the number of elementary cells, but the visualization process is simpler on tetrahedral cells than on the more complex hexahedral ones;

- handling degenerated cases (i.e. coincident points in 3D space) is simplified; tetrahedra degenerates to non-solid cells (areal or linear) and therefore such elements can be simply ignored; on the other hand, the degeneration of hexahedral cells may produce solid cells that have to be accurately managed;

- both data and visualization processes can be simplified by using, for example, a multiresolution representation [1].

Direct projection of tetrahedral cells, using the hardware capabilities of current state of the art graphics workstations, is an efficient process (nearly in the order of $O(10K \div 100K)$ tetrahedral cells per second). Nevertheless, the performance required for the interactive use of these techniques is still far beyond current speeds, especially in the case of low or medium capacity workstations. In this paper we thus evaluate the possible optimizations of the Projected Tetrahedra algorithm [6], i.e. approximations of the rendering process and alternative sorting algorithms, together with the refinement of one of these approaches. In addition, we show that a data simplification approach can also be applied to produce significant speedups while maintaining similar approximations in the images produced. Therefore, we prove William's intuition [8] that true real-time interactive projection can be only obtained through data reduction.

The paper is organized as follows. Projective techniques for the direct visualization of volume datasets are briefly introduced in the next section. Section 3 describes the possible optimization of the Projective Tetrahedra algorithm. In Section 4 we show how a data simplification approach can give better results than the optimization of the "rendering process" presented in the previous section. Our conclusions are drawn in the last section.

2 Projective Direct Volume Rendering

Direct visualization of volume datasets represented by a tetrahedral decomposition was originally proposed by Max et al. [4], and Shirley and Tuchman [6]. Both these solutions apply a density cloud model [11] and solve the visualization task with a two–step approach: tetrahedra are first depth–sorted, then each cell is projected onto the screen, and its color and opacity contributions are composed with the previous ones to form the resulting image. The two proposals differ on the method used to compute the contribution of each cell.

The algorithm by Max et al. implements a software scan conversion process in which the contribution of each cell to the generic pixel is computed by analytically integrating color and opacity along the line of sight.

Shirley and Tuchman's technique, called **Projected Tetrahedra (PT)**, approximates the contribution of each cell with a set of partially transparent triangles. This polygon–oriented approach is faster than the previous pixel–oriented one because conventional graphic hardware can be exploited.

Solutions for the depth sort of tetrahedral grids have been proposed by Williams (*topological* sort on curvilinear grids [8]), and by Max et al. (*numeric distance* sort on Delaunay triangulations [4]).

In this paper we limit ourselves to the projective visualization of datasets represented by Delaunay triangulations[1]. We assume that in the case of a hexahedral cells dataset, it would be decomposed into tetrahedral cells in a preprocessing step. The Projected Tetrahedra algorithm can be briefly described as follows.

PT algorithm:

1. apply a transfer function to each vertex in the dataset to map its scalar value into a color and a density value;

2. sort in depth the cells, back to front, according to the current view (parallel or perspective);

3. for each cell:

    (a) classify the cell according to its projected silhouette (Figure 2);

    (b) decompose the projected silhouette of the cell into triangles (Figure 2);

    (c) find color and opacity values for the *non zero thickness* (NZT) vertex using integration in depth on the cell in world coordinates;

    (d) for each split triangle: scan convert, interpolate color and opacity, compose with the current image (using graphic hardware features for maximal efficiency).

3   Optimization of Projected Tetrahedra

The **PT** algorithm is composed of two independent and sequential phases. We review and evaluate the two possible solutions to the *depth sort* of Delaunay tessellations, and then the optimization of the "per tetrahedra" *projection and composition* phase.

3.1   Depth sort optimization

The so called depth (or visibility) sort must guarantee that the cells are ordered in a such way that, given a viewpoint, if cell a obstructs cell b then b precedes a in the ordering. The two depth sort algorithms proposed in literature for triangulated datasets are the **Meshed Polyhedra Visibility Ordering (MPVO)** algorithm by Williams [9], and the sorting algorithm proposed by Max et al. [4], called in the sequel **numeric distance sorting**.

MPVO is much more general than the second solution because it sorts the cells of any acyclic convex set (another solution for the visibility sort of unstructured meshes has been recently proposed by Stein et al. [7]; it is more general than MPVO, because

---

[1]The Delaunay triangulations of a set of points P in E^3 consists of exactly those tetrahedra with vertices in P which satisfy the property that their circumsphere contains no other points of P into its interior [5].

Dataset					Topolog.		Num.	
	vertices	tetra	T_{adj}	T_{link}	T_{top_sort}	T_{dist}	T_{Qsort}	T_{nd_sort}
BuckyBall								
error 0%	32,768	176,687	41.14	2.09	1.13	0.18	0.78	0.96
error 2%	10,808	63,649	13.67	0.68	0.35	0.07	0.24	0.31
error 5%	6,010	35,909	7.62	0.38	0.19	0.04	0.12	0.16
error 10%	3,088	18,567	3.78	0.18	0.09	0.02	0.06	0.08
BluntFin								
error 2%	2,279	13,094	2.59	0.12	0.08	0.02	0.04	0.06
error 4%	1,012	5,615	1.08	0.05	0.03	0.01	0.02	0.03

Table 1: Resolution of the test datasets and times, in seconds.

allowes multigrid data sorting, but is also more complex in time). The MPVO algorithm requires linear time and uses linear storage. It is a topological solution, based on a pre-processing step (view independent) and two view–dependent steps:

1. an adjacency graph is built, with a node for each cell and a link for each couple of adjacent cells (preprocessing step);

2. links are oriented: each link points toward the obstructing cell, according to the current view; Williams calls this relation *behind* relation [9];

3. a topological sort of the graph, starting from the cells which do not obstruct any other cell, returns a correct visible order.

The **behind** relation is evaluated, for each couple of adjacent cells, taking into account the shared face and the position of the viewpoint. The shared face defines a plane and two halfspaces. The link is therefore directed toward the cell whose half–space contains the viewpoint. To implement this, the plane equation for the shared face has to be evaluated at the viewpoint. The plane coefficients are computed and stored in the preprocessing phase (together with the adjacency graph).

The MPVO algorithm has a linear cost, but requires substantial geometric tests: when the viewpoint changes, all of the adjacency link orientations have to be fixed (i.e., an evaluation of a linear equation for each link).

The cost of MPVO has been empirically evaluated on the BluntFin [2] and the Bucky-Ball [3] datasets, represented using a multiresolution scheme [1]. The results are reported in Table 1, where: T_{adj} is the time for the construction of the adjacency graph (view independent, thus preprocessing), T_{link} is the time for the orientation of the adjacency links, and T_{top_sort} is the time for the topological sort. Times are in cpu seconds on an SGI Indigo R4000 XS24 workstation.

Max et al. reported [4] that, given a Delaunay triangulation and a viewpoint, a visibility ordering is always defined. In addition, reporting results from Edelsbrunner and Joe, they proved that a depth sort can be simply computed without having to manage topological data.

---

[2] Produced and distributed by NASA–Ames Research Center; it is defined on a curvilinear grid composed of $40 \times 32 \times 32 = 40,960$ samples.

[3] Courtesy of AVS International Center; it represents the electron density around a molecule of C_{60} and is defined on a $32 \times 32 \times 32$ regular grid.

This alternative solution, here called **numeric distance sort**, computes for each tetrahedron t_i :

1. the center c_i and the radius r_i of the sphere circumscribed to t_i (such data are view independent, and are thus computed in a preprocessing phase);

2. the square of the distance d_i from the viewpoint to each center c_i;

3. the square length of the segment starting from the viewpoint and tangent to the sphere circumscribed to t_i, i.e. $l_i^2 = d_i^2 - r_i^2$;

It has been proved [4] that the visibility sort corresponds to the sort of segments l_i with respect to their length: a cell t_1 obstructs t_2 *IFF* $l_1 < l_2$. To compute visibility using this approach we need:

preprocessing – to compute and store centers and radii of nt circumspheres;

for each view – to compute $2 * nt$ distances and to sort nt distance values;

where nt is the number of tetrahedra in the triangulation.

In the rightmost columns in Table 1 we report the times required to sort the datasets with the *numeric distance* sort. The total time, T_{nd_sort}, is split into the two subphase times: T_{dist} is the cost of computing distances, and T_{Qsort} is the cost of sorting these distances using an optimized version of the Quicksort algorithm.

A comparison of the two sorting approaches can be driven by the evaluation of the implementation complexity, numerical instability, and efficiency.

The *implementation* of the second solution is straightforward; it does not involve visiting a DAG, and only requires the evaluation of simple mathematical functions and the use of a numeric sorter.

Numerical instability, extremely probable on high resolution datasets, is greatly reduced using the second solution. Obviously, imprecise distances can be computed but this does not halt the sorting process. This problem might at most lead to the incorrect depth ordering of some tetrahedra.

On the other hand, if the first approach is adopted precision is highly critical because errors in the orientation of the links may result in detecting false cycles. More dramatically, in some cases an incorrect ordering of one or more links can prevent an entire region in the grid from being sorted, as is shown in the 2D example in Figure 1.

To evaluate the *efficiency* of the two sorting solutions, we must take into account that the graph orientation is not required to sort the data by the *numeric distance* sorter, but this orientation is needed in the subsequent classification of tetrahedra in projection space (see Subsection 3.2). Therefore, this processing is mandatory whichever sorting strategy is chosen, and its cost is thus not included in the T_{top_sort} times reported in Table 1.

The analysis of the results shows that both for efficiency reasons and higher numerical stability the *numeric distance* sort is the correct choice.

3.2 Projection (classification + rendering) optimization

Once the cells have been depth–sorted, standard PT operates on each cell as follows. Each cell is projected in screen space and its projected silhouette is classified according

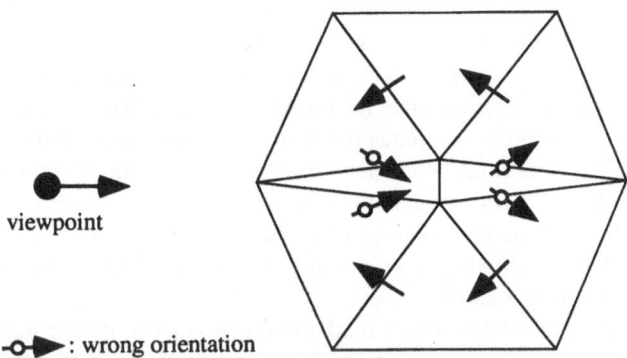

viewpoint

-o▶ : wrong orientation

Figure 1: Four erroneous link orientations generate a region which cannot be sorted.

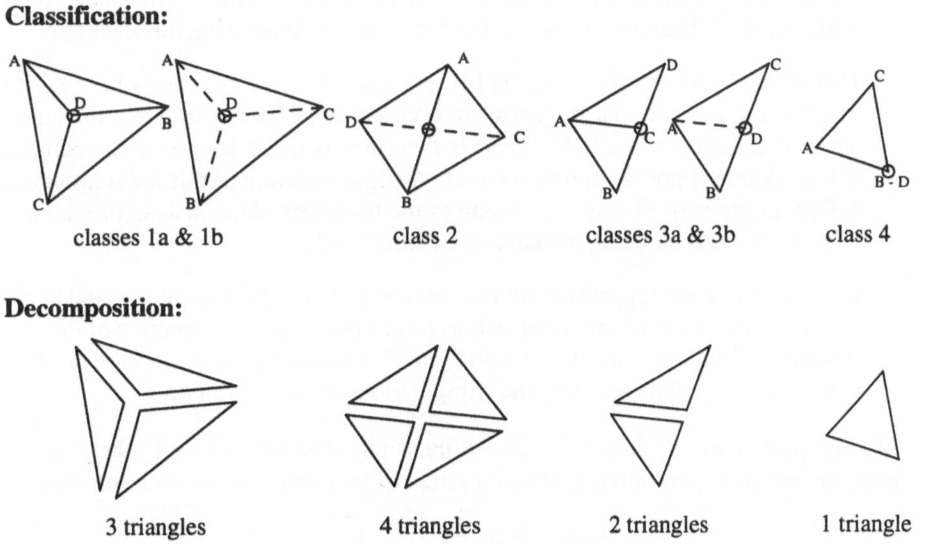

Classification:

classes 1a & 1b   class 2   classes 3a & 3b   class 4

Decomposition:

3 triangles   4 triangles   2 triangles   1 triangle

Figure 2: Classification and decomposition into triangles of a tetrahedral cell depending on the current profile in viewing space (o: NZT vertex).

to Figure 2. Only four projections are possible: two of them which are combinatorially different plus two degenerated cases. The frequency of the degenerated classes 3 and 4 was estimated in less than 5% of the total [10], and therefore a faster management of classes 1 and 2 is taken into consideration in most approximations. Classification of cells directly derives from the orientation of the adjacency links.

For each class, a *non zero thickness* (**NZT**) vertex is determined. The NZT vertex (in screen space) is choosen such that the thickness of the cell in view space is maximal. The standard PT algorithm computes the density and the color of the NZT vertex by evaluating the thickness of the cell in correspondence to the line of sight passing through it. A particle volume density model [11] is adopted.

The projected silhouette is then split into triangles (Figure 2). Each triangle is rendered and blended in hardware as a Gouraud–shaded facet, with a non zero opacity assigned to the NZT vertex alone.

Stein et al. [7] proved that the PT method may produce artifacts, and proposed a solution based on the use of hardware assisted texture mapping. For the sake of simplicity, here we will consider the standard PT method output as *the* correct image, without considering possible artifacts.

There are two possible strategies which can reduce the cost of the projection phase: reducing the number of scan–converted facets and/or avoiding the color and opacity computations for the NZT vertices. With these strategies in mind, the optimization techniques evaluated here are:

- **Voxel** [8]: in this very rough approximation, color and opacity (irrespective of the cell thickness) are precomputed for each cell as an average of the values of its vertices; at projection time, all visible faces are rendered using flat shading;

- **Uniform Thickness Slab** (**UTS1**) [8]: color and opacity are interpolated on the projected silhouette, but the cells are considered as having uniform thickness. The computation of the NZT vertex is therefore avoided, and the corresponding cell thickness is not taken into account. A slight variation of this approximation, **UTS2**, in the case of class 1a visualizes the back face alone instead of the three front faces, thus reducing the number of split facets;

- **Centroid**: this new approximation avoids computing NZT coordinates and thickness by using the centroid of the cell as a splitting vertex, irrespective of the current view. Centroids data (coordinates, color and opacity) are computed in a preprocessing step, thus reducing the computations needed at run time.

To compute centroids data, for each cell the center c of the inscribed sphere and its radius r are computed by solving a linear system of four equations in four unknowns:

$$-\frac{a_i x_c + b_i y_c + c_i z_c + d_i}{\sqrt{a_i^2 + b_i^2 + c_i^2}} = r \tag{1}$$

where the coefficients a_i, b_i, c_i, d_i are the coefficients of the plane on which face i of the tetrahedral cell lies. The scalar value of the centroid is then computed by a simple linear interpolation.

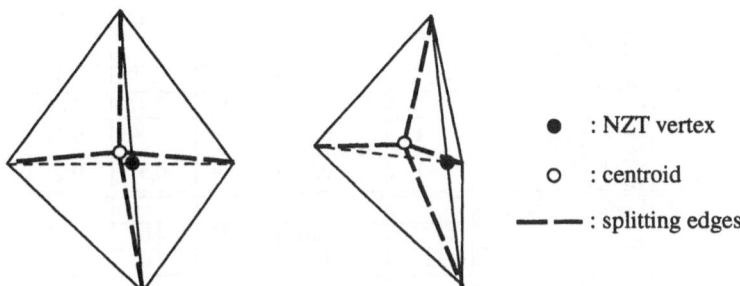

Figure 3: Different projections of class 2 cells.

Centroid approximation manages the splitting phase as follows:

- **class 4:** the same as the PT algorithm;

- **class 3:** the same as the PT algorithm, but the cell thickness is set to $c_1 * r$, where c_1 is an empirical constant;

- **class 2:** the splitting vertex and value are in this case the centroid's coordinates and value; the cell thickness is set to $c_1 * r$;

- **class 1:** the coordinates of the splitting vertex are the coordinates of the NZT vertex (they are combinatorially known, no geometric computations are required). The value of the splitting vertex is set to the mean value between the centroid and the NZT vertex values. The cell thickness is set to $c_1 * r$.

The Centroid approximation therefore avoids the computation of:

- one edge intersection and two color&opacity interpolation for the cells in class 2;

- one plane–ray intersection and one color&opacity interpolation for the cells in class 1;

- one bilinear interpolation and one color&opacity interpolation for the cells in class 3.

Our experiments report that the best approximations were obtained using values of $c_1 = 2.5 \div 3.5$.

Obviously, the approximation given by the centroid method is dependent, for each cell, on the distance between the position of the centroid and the current view–dependent position of the NZT vertex (Figure 3). In order to reduce this approximation error, the opacity of the vertices on the boundary of the projected silhouette was set to 0.5 times the value of the splitting vertex opacity.

When applied to faces which are nearly parallel to the line of sight the method of the centroid may therefore result in a poor approximation (e.g., the rightmost cell in Figure 3). This may be common in the case of datasets obtained by decomposing a regular or curvilinear dataset into tetrahedra, but is much less frequent when the tetrahedral

	PT	Centroid	UTS1	UTS2	Voxel
Bluntfin 2% error					
T_{render}	0.73	0.59	0.36	0.29	0.32
T_{vis}	0.91	0.77	0.54	0.47	0.50
$\frac{T_{render}(approx.)}{T_{render}(PT)}$	1.	0.80	0.49	0.39	0.43
$\frac{T_{vis}(approx.)}{T_{vis}(PT)}$	1.	0.84	0.59	0.51	0.54
BuckyBall 0% error					
T_{render}	12.19	8.67	5.02	4.02	4.41
T_{vis}	15.24	11.64	8.07	7.07	7.46
$\frac{T_{render}(approx.)}{T_{render}(PT)}$	1.	0.71	0.41	0.32	0.36
$\frac{T_{vis}(approx.)}{T_{vis}(PT)}$	1.	0.76	0.52	0.46	0.48

Table 2: Times on two datasets using PT and the evaluated approximations (times in seconds).

dataset is obtained from unstructured data or from the simplification of a curvilinear or regular dataset [1].

Our comparison of the standard PT algorithm and the three approximations considers both processing times and image quality.

The processing times are reported in Table 2, where T_{render} is the time required to project, scan convert and compose the tetrahedra, and T_{vis} is the total time ($T_{vis} = T_{link} + T_{nd\_sort} + T_{render}$, thus all but the preprocessing T_{adj}; for the other results see Table 1). Centroid times were 60% the PT times, but about 30% higher than with the UTS2 method.

An evaluation of the rendering quality can be based on a subjective comparison (a visual analysis) or on a more objective one, based on the computation of differences between the images. The results of objective numeric comparisons are reported in Table 3. Images are compared by computing, for each corresponding pixel pair, the square length of the vector difference:

$$diff(pix_1, pix_2) = \sqrt{(r_1 - r_2)^2 + (g_1 - g_2)^2 + (b_1 - b_2)^2}.$$

The "per pixel" mean value of the difference is reported in Table 3 (with pixels colors represented in the range 0..255). The graph in Figure 4 represents the distribution of the error as a function of the norm of pixel colors in the difference images (only a subrange of the normalized interval [0. .. 1.] is plotted on the X axes).

Visual results are reported in the color plates: in Figures 7, 8, 9 and 10 we show the results of classical PT method and of Centroid, UTS2 and Voxel approximations on the BuckyBall dataset.

To evaluate rendering quality we must differentiate between data obtained by the decomposition of regular hexahedral cells or by the triangulation of irregularly distributed pointsets. On the first datasets, UTS2 approximation gives better images in shorter times. The BuckyBall results in Table 3 seem to be in contradiction with this assertion but in this case, although the global error of Centroid is lower than that of UTS2, the Centroid error is often located in restricted areas and therefore is more evident. On the second class of datasets, the insensitivity of UTS2 to the thickness of the cells makes

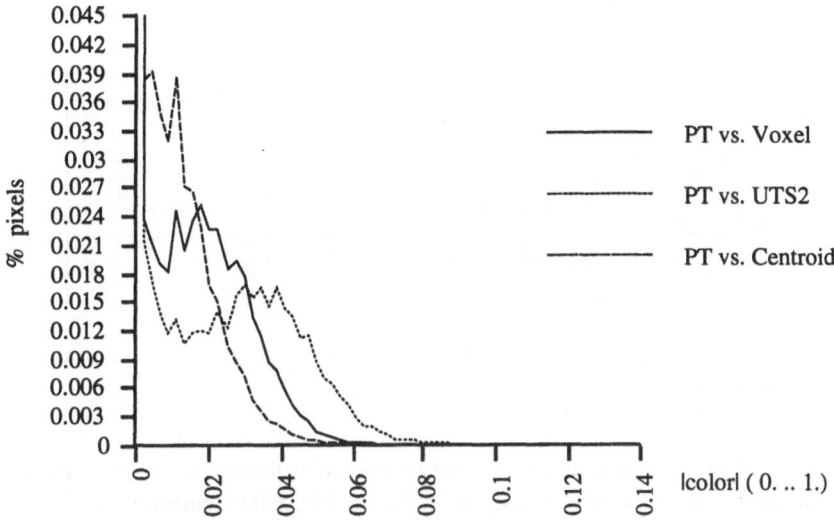

Figure 4: Error distribution in the difference images: standard PT on Buckyball dataset *vs.* approximated methods.

	Mean difference
Bluntfin (2% error)	
PT vs. UTS2	14.65
PT vs. Centroid	10.85
BuckyBall (0% error)	
PT vs. UTS2	8.74
PT vs. Voxel	6.05
PT vs. Centroid	4.33

Table 3: Numerical comparison of the images (standard PT algorithm, UTS2 and Centroid approximations).

Centroid the method that gives better visual and numerical results, though if at the cost of longer processing.

4 Optimization via data simplification

A multiresolution representation for volume datasets has been proposed in a previous paper [1]. Multiresolution is achieved through a sequence of tetrahedralizations approximating the scalar field at increasing precision. The construction of the model is based on an adaptive incremental triangulation approach driven by the local coherence of the scalar field. The result is a number of Delaunay triangulations (i.e. a pyramidal structure) in which each level i gives an approximation of the volume dataset with an error not exceeding a threshold ϵ_i.

Obviously, lower resolution levels may be rendered when faster visualization is re-

Dataset		Times		Speedup	Image Error
	no. tetra	Preproc.	T_{vis}		(mean value)
BuckyBall					
error 0%	176,687	41.14	15.24	1.	0.
error 2%	63,649	13.67	5.02	3.03	2.26
error 5%	35,909	7.62	2.78	5.48	5.29
error 10%	18,567	3.78	1.42	10.73	10.73
BluntFin					
error 2%	13,094	2.59	0.91	1.	0.
error 4%	5,615	1.08	0.39	2.33	3.35
error 6%	2,796	0.52	0.21	4.33	5.39

Table 4: Visualization times obtained using a multiresolution representation of the test datasets (times in seconds).

quired, e.g. for setting visualization parameters such as the current view or the transfer function used to map the field scalar values into colors and densities.

Here, we have experimented with a multiresolution representation to speedup the projection process and evaluate the quality of the images generated.

In order to estimate the performances of the rendering process based on the simplification of the data, we ran the PT code on a number of simplified representations of the two test datasets, Bluntfin and BuckyBall. The processing times reported in Table 4 are the preprocessing time (DAG creation) and the total visualization time T_{vis} (DAG orientation, numeric distance sort, visualization via PT).

Two images produced using the standard PT are reported in Figures 11 and 12; the images are computed on the BuckyBall at precision 2% and 5%.

The graph in Figure 5 reports the distribution of the error over three difference images; for each reduced resolution tetrahedralization (with error 2%, 5%, 10%) the difference image has been computed by subtracting the image rendered on the reduced data from the full resolution Buckyball dataset image (rendered with standard PT method). In the graph, on the abscissa we have the norm of the difference colors (normalized in the interval [0. .. 1.]); on the Y axes are the pixels percentages.

By comparing the results in Tables 2, 3 and 4 it can be noted that images with approximations similar to those obtained using UTS2 are now produced using datasets with a substantial simplification (e.g. on BuckyBall, a similar approximation is obtained using the multiresolution level with a 10% error; see the graph in Figure 6), and therefore with much shorter processing times: about 7 sec. for the UTS2 images and 1.4 sec. for an image using the multiresolution representation at 10% error.

5 Conclusions

In this paper we have focused on the projective visualization of datasets represented via tetrahedral tessellations. Speedup techniques have been reviewed and a new approximation method has been proposed. Approximation methods have been compared in terms of both processing times and the quality of the images produced.

In addition, the results obtained by the optimization of the rendering process have

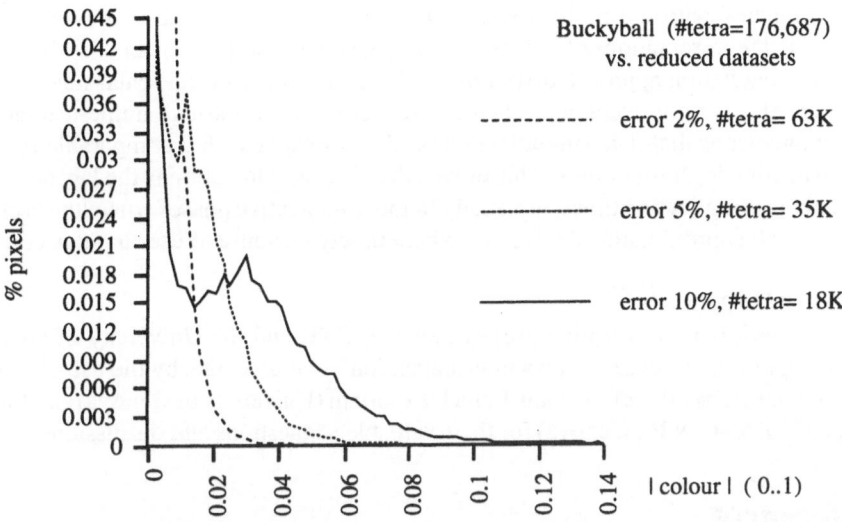

Figure 5: Error distribution in the difference images: standard PT on Buckyball original dataset *vs.* reduced dataset (error 2%, 5%, 10%).

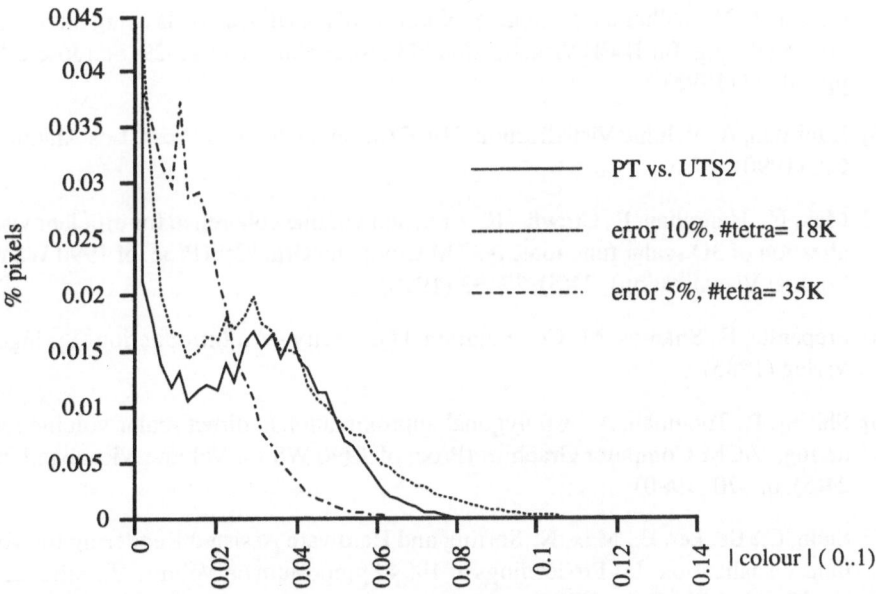

Figure 6: Error distribution in the difference images: comparison of error produced using a rendering process optimization (UTS2) *vs.* data reduction, on Buckyball dataset.

been compared with those obtained by adopting a data simplification approach. Approximated representations of the test datasets have been built using an adaptive incremental triangulation approach driven by the local coherence of the scalar field and are represented in a multiresolution scheme. We have showed through numerical and visual comparisons that data simplification produces images with a comparable level of approximation in shorter times. This proves that data simplification is the key approach to reducing visualization times, especially in those interactive phases where the user tolerates a high approximation degree and where timely response to user input is crucial.

Acknowledgments

This work is part of a joint project between CNR and the University of Genova, which regards the representation and visualization of volume data by the use of simplicial complexes. We therefore thank Leila De Floriani (University of Genova) and Enrico Puppo (I.M.A.–C.N.R., Genova) for their valuable suggestions and discussions.

References

[1] Cignoni, P., Floriani, L. D., Montani, C., Puppo, E., Scopigno, R. Multiresolution Modeling and Rendering of Volume Data based on Simplicial Complexes. In: Proceedings of 1994 Symposium on Volume Visualization, pp. 19–26. ACM Press (1994).

[2] Gelder, A. V., Wilhelms, J. Rapid exploration of curvilinear grids using direct volume rendering. In: IEEE Visualization '93 Proceedings (Oct 25-29, San Jose CA), pp. 70–77 (1993).

[3] Kaufman, A. Volume Visualization. IEEE Computer Society Press, Los Alamitos, CA (1990).

[4] Max, N., Hanrahan, P., Crawfis, R. Area and volume coherence for efficient visualization of 3D scalar functions. A.C.M.Computer Graphics (Proc. of 1990 WS on Volume Visualization), 24(5), 27–33 (1990).

[5] Preparata, F., Shamos, M. Computational Geometry - An Introduction. Springer-Verlag (1985).

[6] Shirley, P., Tuchman, A. A polygonal approximation to direct scalar volume rendering. ACM Computer Graphics (Proc. of 1990 WS on Volume Visualization), 24(5), 63–70 (1990).

[7] Stein, C., Becker, B., Max, N. Sorting and Hardware Assisted Rendering for Volume Visualization. In: Proceedings of 1994 Symposium on Volume Visualization, pp. 83–90. ACM Press (1994).

[8] Williams, P. Interactive splatting of nonrectilinear volumes. In: A. Kaufman, G. Nielson (eds.), Visualization '92 Proceedings, pp. 37–45. IEEE Computer Society Press (1992).

[9] Williams, P. Visibility ordering meshed polyhedra. ACM T.O.G., 11(2), 103–126 (1992).

[10] Williams, P. Interactive Direct Volume Rendering of Curvilinear and Unstructured Data. Ph.D. thesis, University of Illinois at Urbana–Champaign (1993).

[11] Williams, P., Max, N. A volume density optical model. In: A. Kaufman, W. Lorensen (eds.), 1992 Workshop on Volume Visualization, pp. 61–68. ACM Press (1992).

Editors' Note: see Appendix, p. 154 for coloured figures of this paper

Direct Volume Rendering of Irregular Samples

René T. Rau* and Wolfgang Straßer

WSI/GRIS, Universität Tübingen, Auf der Morgenstelle 10,
D-72076 Tübingen, Germany.

Abstract. Visualization of concentrations and density values is one of the main tasks of volume rendering systems. Whenever the sample points are not located on any kind of structured grid most visualization tools are not able to display the data without expensive resampling. We show that a simple forward mapping algorithm can handle this problem efficiently, whenever the reconstruction of the volume function uses spherical kernels. Different examples are discussed and we visualized a simulation from astrophysics which is based on smoothed particle hydrodynamics. Here gas dynamical processes are modelled by a system of pseudo particles where the positions are irregular and vary strongly in time. We produced high quality images which display the simulation data correctly and showed that in this situation our approach is superior to resampling strategies.

1 Introduction

Various fields of sciences produce three dimensional data which need to be visualized. The data produced can have various structures. Besides Cartesian grids where all the cells are identical axis-aligned cubes there exist other structered data on regular, rectilinear, structured and unstructured grids. Finally there exist data where the values are assigned to scattered points in the volume (cf. [1]).
Scattered data are produced from measurements in, e.g., geophysics and chemistry. Smoothed particle hydrodynamics (SPH) simulations are used in the field of astrophysics to treat gas dynamical phenomena. Also this kind of simulation produce data with no underlying grid since it is a particle based model.
In this paper we are not interested in a general scattered data interpolation problem (cf. [4] for a survey). Rather we show that there are many situations, where the reconstruction or approximation uses spherical kernels. In this case we propose a splatting based rendering algorithm which displays the data efficiently.

2 Forward Mapping Algorithm

According to Westover [20] volume rendering is the direct display of data sampled in three dimensions and there are two principle approaches to this: backward mapping algorithms (raycasting) that map the image plane onto the data by shooting rays from each pixel to the scene (cf. [10], [11], [19]), and forward

---

*Supported by the Deutsche Forschungsgemeinschaft, SFB 382

mapping algorithms that map the data onto the image plane (cf. [20] and [12]). Westover restricted his attention to the case of regular grids and developed his so called splatting technique. We start with a more general assumption.

Suppose that a scalar volume function $f : D \subset \mathbb{R}^3 \to \mathbb{R}_+$ can be reconstructed from a finite set of sample values $s_i = s(r_i)$, $r_i \in D$, $i = 0, \ldots, N$ by the equation

$$f(r) = \sum_{i=0}^{N} s_i W(r, r_i),\qquad(1)$$

where W denotes a suitable reconstruction kernel.

For the visualization of this volume function a general model based on the stationary linear transport equation can be used. This model was described by Krueger (see. [9] for details) and treats most common rendering models as special cases.

A simple transport equation can be formulated in the situation where no scattering and no absorption is present. The solution Φ of the transport equation is then given by the line-of-sight integral

$$\Phi(r, \Omega) = \int_0^{\infty} f(r - R\Omega)\, dR,\qquad(2)$$

where Ω denotes a direction, R a real parameter and f is used as the source term for the virtual particles. For parallel projection this equation has to be evaluated for every point r in the viewplane and a fixed viewing direction $-\Omega$. The benefit of this model was desribed in, e.g., [17].

In the situation given by Equation (1) we can interchange integration and summation and obtain

$$\Phi(r, \Omega) = \sum_i s_i \int_0^{\infty} W(r - R\Omega, r_i)\, dR.\qquad(3)$$

The main idea of Westover was to compute in a preprocessing step the integration and then to compute only the summation within the rendering step.

In our more general situation we have to ask first for suitable kernels, where the approach of Westover can be implemented efficiently in comparison to resampling.

3 Suitable Kernels

The advantage of resampling the given function on a regular grid is that on these structures we can use very efficient algorithms to compute the line-of-sight integral given by Equation (2), e.g., raycasting algorithms. Since resampling can be done in a preprocessing step we would have no delay whenever we want to display different viewpoints. On the other hand the main problem with respect to resampling is aliasing whenever the function is not band-limited or the sample frequency is not high enough. Also the determination of the sample rate can be difficult. Additionally high sample rates yield large grids.

In our approach with separation of integration and summation the disadvantages of resampling do not appear. Nevertheless the kernels should satisfy several conditions in order to obtain reasonable rendering time.

One important property the kernel should satisfy is the translation invariance, i.e.,

$$W(r, r') = W(r - r') \tag{4}$$

Whenever this condition is fulfilled and

$$\int_0^\infty W(r - r_i - R\Omega)\,dR \approx \int_{-\infty}^\infty W(r - r_i - R\Omega)\,dR \tag{5}$$

holds we can rewrite Equation (3) and obtain

$$\Phi(r, \Omega) = \sum_i s_i I_\Omega(r - r_i). \tag{6}$$

where $I_\Omega(r') = \int_{-\infty}^\infty W(r' - R\Omega)\,dR$. Thus we can evaluate the integral independent of the sample location.

We remark that Relation (5) is automatically fulfilled if the domain D lies within the viewing volume and W decreases "rapidly" as a function of the distance to the viewing plane.

If the volume is decomposed into tetrahedra with, e.g., linear interpolation, the assumption (4) will not be fulfilled and other visualization techniques will be appropriate (cf., e.g., [18])

Since resampling of the volume function on a regular grid has the advantage of being viewpoint independent we should have this property in our approach, too. Demanding that the kernel is spherical, i.e.,

$$W(r) = W(|r|),$$

we get $I_\Omega(r) = I(|r|)$ independent of the direction Ω.

Therefore our approach will be competitive against resampling whenever it is possible to use translation invariant, spherical and rapidly decreasing reconstruction kernels.

In the next section we give several examples, where our assumption are satisfied and a splatting-based forward mapping algorithm is reasonable.

4 Examples of Spherical Kernels

Exact Reconstruction

Sampling and reconstruction in 3 dimensional space of band-limited functions have higher degrees of freedom than in the one dimensional situation (cf. [16]). The most efficient sampling lattice is in general not rectangular nor exists a unique reconstruction function for a given lattice (cf. [16]).

For a given lattice for which the repetitive spectra do not overlap, every function which equals some constant Q on the spectrum and is equal to zero whenever

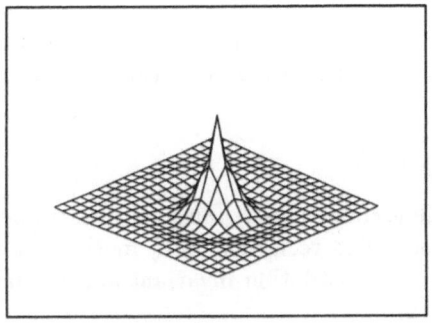

Figure 1: Integration of the kernel given by Equation (6) usefull for exact reconstruction.

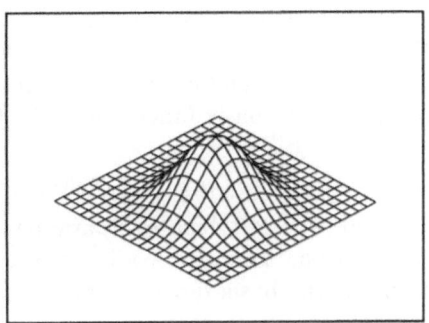

Figure 2: Integration of the kernel given by Equation (10) used in the SPH simulation with $h = 1$.

the repetitive spectra are not vanishing, can be used for exact reconstruction (see [16] for details).

For a compact spectrum σ of a function which is nearly rotational symmetric, i.e. $\sigma \subset K_{2\pi B} = \{x \in \mathbb{R}^3, |x| \leq B\}$, we can use the inverse Fourier transform of the function

$$g = Q \cdot \chi_{K_{2\pi B}},$$

where $\chi_{K_{2\pi B}}$ denotes the characteristic function on $K_{2\pi B}$. The resulting spherical kernel is then given by

$$W(r) = \frac{\pi}{\sqrt{2}(2\pi B|r|)^3}(\sin 2\pi B|r| - 2\pi B|r| \cos 2\pi B|r|). \qquad (7)$$

This kernel satisfies our assumptions. Another important kernel which is often used in reconstruction and satisfies our conditions is of course the three dimensional Gaussian kernel. Other kernels usefull in the field of computer graphics which fit in our framework are described in [13]. We remark that in [13] other advantages of our kernel conditions concerning the image quality are mentioned. In the case of irregular samples the problem of reconstruction of a band-limited function is more complex and several authors investigated this problem in recent years (cf. [2], [6]). In [2] the problem was treated in great generality. They showed that under suitable conditions a given function f can be represented as a series of translates $L_{r_i}g$ for a given integrable, band-limited function g. This means that f can be reconstructed by the formula

$$f(r) = \sum c_i g(r - r_i) \qquad (8)$$

with iteratively computed values c_i. Since the assumptions on g are not very restrictive (even the integrability can be dropped) we can use the spherical kernels given by Equation (7) for the reconstruction.

Scattered Data Interpolation

For scattered data interpolation radial basis function can be used. One of the most successful and most applied method is Hardy's multiquadrics (MQ) method (cf. [4]). Here basis functions of the form

$$g(r) = (|r|^2 + R^2)^{\frac{\mu}{2}}$$

are used, where R denotes a parameter which is choosen depending on the sample distribution. For $\mu = -1$ it leads to the so-called reciprocal MQ method. In this case the basis functions are rotational and translation invariant and are in general rapidly decreasing.

The interpolating function is then given by $f(r) = \sum a_i g(r - r_i)$ where the a_i are obtained by solving a system of linear equations resulting from the interpolation condition.

Function Approximation

Among the various possible approximations of functions we choose the following. Suppose

$$f : D \subset \mathbb{R}^3 \to \mathbb{R}^n \in L^2$$

is given. We approximate f by the kernel estimate

$$< f(r) >= f_h(r) = \int_D W(r, r', h) f(r') \, dr', \tag{9}$$

where W is the so-called *smoothing kernel* and h the *smoothing length* (cf. [14]). The smoothing kernel has to be normalized, i.e. the volume integral of W over its support equals 1

$$\int W(r, r', h) \, dr' = 1.$$

Furthermore W has to be choosen in a proper way such that

$$f_h(r) \xrightarrow{h \to 0} f(r).$$

All this can be performed by kernels which are of the form

$$W(r, r', h) = W(|r - r'|, h).$$

We remark that this kind of representation is often used in SPH simulations. If we consider the values of f only at a finite number of points, distributed with the number density $n(r) = \sum_{i=1}^N \delta(r - r_i)$, we can multiply the integrant in Equation (9) by $n(r')/ < n(r') >$ and obtain the following discrete approximation

$$f(r) \approx< f(r) >\approx \sum_i \frac{f(r_i)}{< n(r_i) >} W(|r - r_i|, h) \tag{10}$$

where $< n(r) >= \sum_i W(|r - r_i|)$ (cf. [7]).

In the special case of mass density ρ necessary in gas dynamical simulations we have to replace the function f by ρ, $< n(r_i) >$ by $\rho(r_i)/m_i$ and obtain the following final approximation

$$\rho(r) = \sum_i m_i W(r - r_i, h)$$

where m_i denotes the mass of the pseudo-particle i (cf. [7]).

We can interpret this final equation in such a way that the mass of the particles is distributed in space according to the values of the kernel.

5 Application: SPH Simulation

Gas dynamical processes play an important role in the evolution of astrophysical systems. In order to verify model assumptions on an astrophysical phenomen scientists are often using numerical calculations. SPH was used recently with great success in this field (cf., e.g., [5] and [3]).

With the approximation procedure mentioned above all interesting thermodynamical quantities can be approximated in the same manner with the same kernel W. Since in most applications the kernel is choosen with compact support the smoothing length also determines the interaction radius of the particles. Obviously the used kernels satisfy our conditions. Moreover our rendering technique is able to display the interaction radius exactly.

The simulation data we visualized have been supplied by Peter Kroll, Theoretische Astrophysik Tübingen. The group of Hanns Ruder investigated the model assumption that Be star disks are formed by ejection of stellar matter from a point source at the equator rotating at critical velocity (cf. [8]).

They used constant mass for the particles, i.e. m_i=const. For the smoothing kernel they used a spline given by

$$W(q) = \begin{cases} \frac{8}{\pi h^3}(1 - 6q^2 + 6q^3) & : \quad 0 \leq q < \frac{1}{2} \\ \frac{16}{\pi h^3}(1 - q)^3 & : \quad \frac{1}{2} \leq q < 1 \\ 0 & : \quad 1 \leq q \end{cases} \tag{11}$$

where $q = |r|/h$. For the simulation they used a smoothing length h of $5 \cdot 10^{10}$ (see Figure 3-6).

The particles are ejected from a point on the surface of the rotating star and therefore the locations of the particles vary strongly in time. Nevertheless our method has to handle only the number of particles in a certain timestep independent of the location. Since the simulation used only about 30000 particles our rendering can be performed very fast. With resampling we should use one single grid for all time steps in order to avoid defects between two frames. Thus for high quality rendering a very large grid has to be used which results in a long rendering time. These arguments show that our forward mapping algorithm is the only reasonable way to display SPH simulations correctly and fast.

6 Implementational Aspects

To allow perspective viewing of the volume data and to be able to render hybrid structures composed out of volume and surface data we used a polygonal based forward mapping technique. The advantage of this approach is that hardware assistance can be used for compositing, i.e. summation (cf. [12], [18] and [21]). So we performed the integration of the kernel in Cartesian coordinates once. The resulting footprint was approximated by a planar triangle mesh consisting of only a few triangles with different transparency values at the vertices according to the values of the integration result. Furthermore we choose a single color for all vertices (see below). In the cases where positive and negative values appear we built two meshes which are treated separately.

To every coordinate value of the samples we then translated this planar mesh with normal orthogonal to the viewing plane.

The summation due to Equation (3) has the advantage of being independent with respect to the ordering of the sample locations. The compositing can be performed using the accumulation buffer. The problem within this approach is that frames have to be composed instead of only adding the triangle meshes.

To overcome this problem we used α-blending. As blending function we used (Source,1-Source) of OpenGL (see [15] for details). This means that, in our situation with a single RGB color, e.g. (1,0,0), the new destination value of the red component will be computed from the α-value α_s of the source and the α-value α_d of the destination color by the formula

$$\alpha_s + (1 - \alpha_s)\alpha_d$$

With this approach the mapping of the scalar values into the color space and the compositing can be done in one single step.

Another important advantage of this kind of compositing is the fact that it is independent of the ordering of the samples. This can be easily checked because we used one single RGB color.

α-blending not exactly performs the summation in Equation (6). It can be shown that for small α-values we have an approximated solution of the transport equation where some small absorption is present.

In the situation where we have to visualize additional surface data in the scene we used the z-buffer algorithm to eliminate hidden surfaces. In this case we had to sort the sample points with respect to the distance to the view plane in order to obtain correct blending results. The same must be done if we want to perform an additional color coding.

For the SPH-simulation we used a net consisting of 8 triangles with a size given by the smoothing length. Without sorting we obtained a rendering time of about 15 seconds for 36000 sample points on an Indy SC. Here no hardware assistance for α-blending was available and a significant acceleration should be possible if α-bitplanes are present.

7 Conclusions

We investigated the direct rendering of volume data where the sample points are scattered and the reconstruction uses kernels. We found that for a simple rendering model the data can be rendered efficiently with a splatting technique, whenever the reconstruction kernel satisfies some properties. These properties are translation invariance, rotational invariance and rapid decrease. Furthermore we showed that there are various examples of kernels where these properties are satisfied.

We investigated the SPH simulations and found that in this case our assumptions are fulfilled and our rendering technique displays exactly the function used in the simulation.

Since our visualization model coincides with the simulation assumptions for SPH we produce relyable images which give the scientist faster and correcter insight in the simulation data as resampling techniques.

References

[1] T. T. Elvins: A survey of algorithms for volume visualization. *Computer Graphics*, 26(3), 194–201 (1992).

[2] H.G. Feichtinger and K. Gröchenig: Iterative reconstruction of multivariate band-limited functions from irregular sampling values. *SIAM J. Math. Anal.*, 23, 244–261 (1992).

[3] O. Flebbe, S. Münzel, H. Herold, H. Riffert, and H. Ruder: Smoothed particle hydrodynamics: Physical viscosity and the simulation of accretion disks. *Preprint* (1994).

[4] R. Franke and G.M. Nielson: Scattered data interpolation and applications: A tutorial and survey. In Geometric Modelling: Methods and Their Applications, H. Hagen & D. Roller (eds.), Springer, 131–160 (1990).

[5] R. A. Gingold and J.J. Monaghan: Smoothed particle hydrodynamics: theory and application to non-spherical stars. *M.N.R.A.S.*, 181, 375–389 (1977).

[6] K. Gröchenig: Reconstruction algorithms in irregular sampling. *Mathematics of Computation*, 181–194 (1992).

[7] L. Hernquist and N. Katz: TreeSPH: A unification of SPH with the hierarchical tree method. *Preprint*, (1994).

[8] P. Kroll, R. W. Hanuschik, H. Riffert, and H. Ruder: 3D-SPH simulations of disks in be stars. *Preprint*, (1994).

[9] W. Krueger: The application of the transport theory to visualization of 3-d scalar data fields. *Computer in Physics*, 397–406 (1991).

[10] Marc Levoy: Display of surfaces from volume data. *IEEE Computer Graphics and Applications*, 8(3), 29–37 (1988).

[11] Marc Levoy: A hybrid ray tracer for rendering polygon and volume data. *IEEE Computer Graphics and Applications*, 10(2), 33–40 (1990).

[12] David Laur and Pat Hanrahan: Hierarchical splatting: A progressive refinement algorithm for volume rendering. In Thomas W. Sederberg, editor, *Computer Graphics (SIGGRAPH '91 Proceedings)*, 25, 285–288 (1991).

[13] Don P. Mitchell and Arun N. Netravali: Reconstruction filters in computer graphics. In John Dill, editor, *Computer Graphics (SIGGRAPH '88 Proceedings)*, 22, 221–228 (1988).

[14] J. J. Monaghan: Why particle methods work. *SIAM J. Sci. Stat. Comput.*, 3(4), 422–433 (1982).

[15] J. Neider: *OpenGL Programming Guide*. Addison-Wesley, 1993.

[16] D.P. Petersen and D. Middleton: Sampling and reconstruction of wave-number-limited functions in n-dimensional euclidean spaces. *Information and Control*, 5, 279–323 (1962).

[17] H. Ruder, T. Ertl, F. Geyer, H. Herold, and U. Kraus: Line-of-sight integration: A powerful tool for visualization of three-dimensional scalar fields. *Computers and Graphics*, 13(2), 223–228 (1989).

[18] Peter Shirley and Allan Tuchman: A polygonal approximation to direct scalar volume rendering. In *Computer Graphics (San Diego Workshop on Volume Visualization)*, 24, 63–70 (1990).

[19] Craig Upson and Michael Keeler: VBUFFER: Visible volume rendering. In John Dill, editor, *Computer Graphics (SIGGRAPH '88 Proceedings)*, 22, 59–64 (1988).

[20] Lee Westover: Footprint evaluation for volume rendering. In Forest Baskett, editor, *Computer Graphics (SIGGRAPH '90 Proceedings)*, 24, 367–376 (1990).

[21] Jane Wilhelms and Allen Van Gelder: A coherent projection approach for direct volume rendering. In Thomas W. Sederberg, editor, *Computer Graphics (SIGGRAPH '91 Proceedings)*, 25, 275–284 (1991).

Editors' Note: see Appendix, p. 155 for coloured figures of this paper

Biorthogonal Wavelet Filters for Frequency Domain Volume Rendering

Roberto Grosso and Thomas Ertl

Lehrstuhl für Graphische Datenverarbeitung (IMMD 9)
Universität Erlangen-Nürnberg
Am Weichselgarten 9, 91058 Erlangen
{grosso,ertl}@informatik.uni-erlangen.de

Abstract.
Rendering images from three-dimensional discrete data sets usually involves interpolation between samples. In terms of signal processing theory, common interpolation methods like trilinear and cubic interpolation are equivalent to the convolution of the sampled data with a suitably chosen reconstruction filter. Frequency domain volume rendering is a technique based on the Fourier projection-slice theorem for the efficient generation of line integral projections without absorption. The quality of the images relies almost completely on the quality of the interpolation filter for the extraction of a 2D slice from the 3D frequency domain representation of the volume. This paper presents experiences we obtained when implementing frequency domain volume rendering and investigates the use of scaling functions of biorthogonal wavelets as reconstruction filters that exhibit the required compact support in space and fast decay in the frequency domain. This method generates X-ray-like images with good quality and short rendering times. In order to accelerate the rendering process without much loss of image quality we introduce wavelets as a subband filtering scheme generating a hierarchical representation of the volume data with the potential for interactive data exploration.

1   Introduction

The problem of visualizing three-dimensional volumetric data is one of the major areas of research in scientific visualization. Although computationally much more expensive than indirect visualization of scalar fields using isosurfaces, direct volume rendering has attracted much attention because of its ability to present the full volumetric structure of the dataset through the use of transparency and various mapping techniques.

If one is concerned with the analysis of standard algorithms for direct volume rendering, it is possible to make some general statements about computational complexity. As in many other areas of computer graphics and visualization there are two broad classes of direct volume rendering methods. Image space methods (e.g. [8]) proceed from pixel to pixel casting rays into the volume to accumulate intensity and opacity along the line-of-sight. Object space methods sort the voxels of the volume in order to traverse them in a front-to-back scheme and successively composite the contribution of the voxel to the pixels in the image [16]. From the point of view of analysis of algorithms and problem complexity, all these *direct* methods have in some sense complexity $O(N^3)$.

A different strategy for overcoming this fundamental problem is based on the three-dimensional version of the Fourier projection-slice theorem [7, 10]. This theorem states that image projections can be obtained by inverse Fourier transforming a 2D slice from the 3D spectrum of the input data. Using fast Fourier transform algorithms (FFT), it can be shown that the complexity of this projection is of order $O(N^2 logN)$. This is, however, a theoretical prediction. In practice one is confronted with the problem of interpolating a sub-manifold from a higher dimensional manifold that is given as sampled data. In our case the samples of the 2D slice do not in general coincide with those of the 3D spectrum. In order to interpolate the function values one first *reconstructs* the function by convolution with a reconstruction filter. The choice of a reconstruction filter can have important consequences on image quality. High-quality interpolation schemes are expensive, thus one usually has to make a trade off between image quality and computing time. In spite of the fact that the inverse FFT is asymptotically dominant compared to the 2D interpolation cost of $O(N^2)$, for standard filters and data sizes interpolation dominates the rendering process. Totsuka et al. [14] have proposed an adaptive algorithm which produces very good results and a performance improvement between 12 - 17% depending on the data properties.

In this paper we present a method for handling the problem of high interpolation cost, which is based on a subband filtering scheme with exact reconstruction using the wavelet transform. After an *encoding* filtering step the input data is transformed into two *signals*, one which is a smooth down-sampled approximation of the input data, obtained using a low-pass filter. The second signal contains the detail information used for the *decoding* to obtain the original data. The down-sampled smooth component contains fundamental information of the data set, which can be used for fast rendering and data exploration purposes. Depending on the approximation properties and on the regularity of the approximating basis of the wavelet transform, images of very good quality can be produced with a performance improvement of factor 4. Retaining only those samples of the detail components that are above a user given threshold, one can take advantage of the wavelet schemes for lossy data compression.

In Section 2 we review the fundamentals of frequency domain volume rendering and present some technical issues that are of practical importance. Section 3 presents some results of the wavelet theory which are relevant for our application. In Section 4 a hierarchical method which allows the user to trade off between performance and image quality is introduced and some results are given in section 5.

2 Frequency Domain Volume Rendering

With respect to medical imaging volume rendering can be regarded as the inverse problem of tomographic reconstruction where the unknown density distribution $f(x, y, z)$ has to be derived from a set of measured projections. Volume rendering, on the other hand, generates images by projecting the reconstructed density distribution onto a virtual screen. Using transfer functions to assign color and opacity to the density values allows for both semi-transparent and opaque surface like effects. Fourier volume rendering [10] is a technique which generates line integral projections of the spatial data for the special case where there is no absorption, thus producing X-ray-like pictures preferred by many physicians. The advantage of this approach is that the Fourier projection-slice theorem allows us to compute 2D projections of 3D data by interpolating a 2D slice of the data representation in the frequency domain and transforming it back into the spatial domain which can be done in $N^2 \log N$ time for a scalar array of size N^3.

2.1 Theoretical Background

If $f(\vec{x})$ denotes the density distribution, the X-ray projection shadow along a ray is proportional by a factor of k to the line integral of this density along the ray. Considering a ray that intersects the image plane spanned by the unit vectors $\vec{n_1}$ and $\vec{n_2}$ at point p with plane coordinates u, v the line integral takes the form:

$$p(u, v) = k \int_{R^3} f(\vec{x}) \delta(\vec{x} \cdot \vec{n}_1 - u) \delta(\vec{x} \cdot \vec{n}_2 - v) \mathrm{d}\vec{x}$$

Using this expression the Fourier transform $P(\xi, \zeta)$ of $p(u, v)$ can be computed:

$$P(\xi, \zeta) = k \int_{R^2} \int_{R^3} f(\vec{x}) e^{-i(\xi u + \zeta v)} \delta(\vec{x} \cdot \vec{n}_1 - u) \delta(\vec{x} \cdot \vec{n}_2 - v) \mathrm{d}\vec{x} \mathrm{d}u \mathrm{d}v$$

Interchanging the order of integration and solving the integrals results in the 3D slice-projection theorem, which states that the Fourier transform of p can be obtained by evaluating the Fourier-Transform $F(\xi, \zeta, \eta)$ of $f(\vec{x})$ at a plane parallel to the image plane:

$$P(\xi, \zeta) = F(n_{1x}\xi + n_{2x}\zeta, n_{1y}\xi + n_{2y}\zeta, n_{1z}\xi + n_{2z}\zeta)$$

The final image $p(u, v)$ can be obtained by computing the inverse Fourier transform of $P(\xi, \zeta)$.

Following [14] this technique will be refered to in this paper as frequency domain volume rendering. The main motivation for this approach is that, once the forward 3-D transform is computed in a preprocessing step, we can compute projections at arbitrary angles by interpolating 2D data sets in the frequency domain and back transforming them. The projection-slice theorem presented above is based on some general results of transform analysis and therefore also holds for the Hartley transform [2] which will be used throughout the rest of the paper, because it generates real values from real input data sets whereas the Fourier transform is likely to produce complex values.

Summarizing, the base algorithm for frequency domain volume rendering consist of the following steps: **1)** compute the 3D spectrum of the input data by the Hartley transform as a preprocessing step, **2)** for a given viewing direction interpolate a 2D slice orthogonal to this direction from the 3D spectrum, **3)** obtain the final projection image by applying the inverse Hartley transform to this 2D slice.

2.2 Practical Issues

Obviously, the theoretical considerations presented above dealt with continuous functions and their continuous projection images. In practical applications, however, we are given volumes of discrete samples and we are concerned with the computation of discrete color or grey level images. Therefore, when performing the volume rendering by integrating along the line-of-sight, the function values have to be interpolated from the data samples, making assumptions on the nature of the continuous function that is being modeled by the discrete samples. Traditional volume rendering algorithms are either based on the voxel model, introducing discontinuities between voxels, on trilinear interpolation, which leads to discontinuities in the first derivatives, or on quadratic interpolation with discontinuous second derivatives.

Considering that we are concerned with data that will be subject to the discrete Fourier transform, we assume that our density distribution function $f(x, y, z)$ is a band limited function whose highest frequency is given by the spatial sampling rate. In this case, f is infinitely differentiable and completely specified by its samples. With this model in mind, we now discuss some issues of practical interest for the implementation of the algorithm described above.

The Discrete Hartley Transform: When working with digital computers we are constrained to a discrete world and we have to employ the *discrete* Hartley transform. In order to interpret its results correctly, the relation between the discrete and the continuous transform have to be kept in mind. Let us consider a continuous one-dimensional function $f(x)$ and its continuous transform $F(\xi)$. This transformation pair has to be modified using signal processing methods in order to obtain the corresponding discrete transformation pair $\tilde{f}(x)$ and $\tilde{F}(\xi)$, which represents an approximation to the continuous transform. The first step is to sample f with the comb function, obtaining the discrete function $\hat{f} = f(x)\texttt{comb}(x, T)$, where T is the sampling rate. Since we cannot process infinitely many samples, we restrict this function to an interval around a point x_0 using the box-filter $\texttt{rect}(x, x_0, T_0)$, where T_0 is the size of the interval. The continuous Hartley transform of the discretized and restricted $\hat{f}(x)\texttt{rect}(x, x_0, T_0)$ is the periodic function $F(\xi)*\texttt{comb}(\xi, 1/T)*\texttt{sinc}(\xi, 1/T_0)$ with period $1/T$ and the symbol $*$ meaning convolution. The last step is to discretize this function again using $\texttt{comb}(\xi, 1/T_0)$. The discrete transformation pair is given by the periodic continued functions

$$\tilde{f}(x) = f(x)\text{comb}(x, T)\text{rect}(x, T_0) * \text{comb}(x, T_0)$$
$$\tilde{F}(\xi) = F(\xi) * \text{comb}(\xi, 1/T) * \text{sinc}(\xi, 1/T_0)\text{comb}(\xi, 1/T_0).$$

The sinc function is called the reconstruction filter.

Resampling and Reconstruction Filters: Now, we have to deal with the consequences of the fact that we are not actually working with f and F, but with the discrete periodic functions $\tilde{f}$ and $\tilde{F}$. What we are actually looking for is the period of $\tilde{f}$ centered at zero, all other *copies* of this interval are called alias spectra of F. The same is true for F, where we talk about the alias spectra of f. Since we need the values of F between samples, we use a reconstruction filter H which is convolved with the discrete signal: $F(\xi) = \tilde{F} * H(\xi)$. The convolution theorem for the Hartley transform states that the transform of $\tilde{F} * H(\xi)$ is

$$\frac{1}{2}\tilde{f}(x)h(x) - \tilde{f}(-x)h(-x) + \tilde{f}(x)h(-x) + \tilde{f}(-x)h(x)$$

which reduces to the well known form $\tilde{f}(x)h(x)$ as known from the Fourier transform, if the reconstruction filter h is an even function. Considering that the Hartley transform of the sinc function is the rect function and following Shannon's sampling theorem, it can be stated that an exact reconstruction of F is possible, if we use the sinc function as a reconstruction filter. In this case the rect function will cut off the alias spectra of F. The corresponding *Nyquist frequency* and *Nyquist region* are clearly determined by the sampling rate of the discrete Hartley transform [3]. These results will carry over to the three-dimensional case where the sinc and rect functions are just the tensor products of their one dimensional versions.

Compact Support Reconstruction Filters: The sinc function, however, has infinite support and cannot be used as a reconstruction filter in effective computations. A common technique to handle this problem is called data windowing, which means that the sinc function is clipped against a *window*. Many different windowing functions can be found in the literature such as the Welch window, the Hanning window and the Hamming window, which is a Hanning window not becoming exactly zero at the ends, and there are many more elaborated techniques for designing reconstruction filters, e.g. a technique called Projection on Convex Sets (POCS) [5]. In our implementation of the frequency domain volume rendering algorithm we used the Hamming window as suggested by Malzbender [10] which produces quite good results.

Imperfect Reconstruction: If the reconstruction filter has compact support, then its frequency response will have infinite extent, a fact which is called *uncertainty principle* in the signal processing community. Its consequence is that using filters of compact support for resampling means that exact reconstruction is not possible any more. Defects arising from imperfect reconstruction were discussed by Marschner and Lobb [11]. The fall-off of the filter response beyond the Nyquist frequencies is called the *leakage width* and is responsible for aliasing effects because some energy of the spectrum leaks through into the reconstruction.

86

What actually happens is that one will see copies of the periodically extended discrete function which is being modeled. The leakage width is inversely proportional to the filter width, which states that aliasing can be reduced by using a wide filter, but it cannot not be eliminated completely. The importance of good reconstruction filters when working with discrete data sets should have become obvious by now. Filter design will always be a trade-off between the width of the filter and the leakage tails of its response. Large filters will produce good reconstruction but come at high cost because performance degrades when the interpolation of a function value is based on many data samples in a large neighborhood.

3 Wavelets

The importance of orthonormal and biorthogonal basis of wavelets and Multiresolution Analysis is its hierarchical nature, which offers a mathematical framework for describing functions at different levels of resolution. This is a fundamental concept when looking for hierarchical algorithms or hierarchical data manipulation. This section discusses those aspects relevant to our work. Details on wavelet theory can be found in [6, 4].

3.1 Orthonormal Wavelets and Multiresolution Analysis

A multiresolution analysis can be thought of as a ladder of approximating closed subspaces $(V_j)_{j\epsilon Z}$ of $L^2(\Re)$ which satisfies the following conditions:

$$\cdots V_1 \subset V_0 \subset V_{-1} \subset \cdots$$

$$\bigcup_{j\epsilon Z} V_j = L^2(\Re) \quad \text{and} \quad \bigcap_{j\epsilon Z} V_j = \{0\}$$

$$f\epsilon V_j \Leftrightarrow f(2^j.)\epsilon V_0 \quad \text{and} \quad f\epsilon V_0 \Rightarrow f(.-n)\epsilon V_0 \quad \forall n\epsilon Z$$

Furthermore, there exists a function $\phi \epsilon V_0$ such that $\{\phi_{0,n}\}_{n\epsilon Z}$ is an orthonormal basis of V_0, where $\forall j, n\epsilon Z$, $\phi_{j,n} = 2^{j/2}\phi(2^{-j}x - n)$.

The most important result of multiresolution analysis is as follows: given such a ladder of subspaces of $L^2(\Re)$, which satisfies the above conditions, one can construct an orthonormal wavelet basis $\{\psi_{j,n}; j, n \epsilon Z\}$, with $\psi_{j,n} = 2^{j/2}\psi(2^{-j}x-n)$, such that for any function f in $L^2(\Re)$

$$P_{j-1}f = P_j f + \sum_{n\epsilon Z} <f, \psi_{j,n}> \psi_{j,n}, \tag{1}$$

where P_j is the orthogonal projection onto V_j

$$P_j f = \sum_n <f, \phi_{j,n}> \phi_{j,n}.$$

The function ψ is sometimes called the *mother* wavelet. The projection $P_j f$ onto the subspaces V_j corresponds to the different resolution levels in which the function f can be decomposed. These projections contain the *smooth* information of f at a given level of resolution. The projections

$$Q_j f = \sum_{n \in Z} < f, \psi_{j,n} > \psi_{j,n}$$

onto the subspaces W_j spanned by the $\psi_{j,n}$ represent the *detail* information of f required to go from one resolution approximation subspace to the next finer one. Equation (1) is the wavelet decomposition of the function f.

It can be shown from the conditions imposed above on the multiresolution analysis, that the *scaling* function ϕ satisfies the *two-scale* relation

$$\phi = \sum_n h_n \phi_{-1,n} ,$$

which means that the scaling function corresponding to an approximation subspace can be obtained from the scaling function corresponding to the next finer approximation subspace using the discrete *low-pass filter* $\{h_n\}$.

3.2 Filtering Scheme with Exact Reconstruction

There is a simple relation between the wavelet multiresolution analysis and a subband filtering scheme using *quadrature mirror filters* QMF [9]. We start with a scale approximation $f^{j-1} = P_{j-1} f$ of a function f in V_{j-1} and decompose it into a coarser approximation in V_j. Due to the fact that $V_{j-1} = V_j \oplus W_j$, we have $f^{j-1} = f^j + \delta^j$, where $\delta_j = Q_j f$. In terms of the orthonormal bases $\{\phi_{j,n}\}_{n \in Z}$ and $\{\psi_{j,n}\}_{n \in Z}$, we have

$$f^{j-1} = \sum_n c_n^{j-1} \phi_{j-1,n} , \quad f^j = \sum_n c_n^j \phi_{j,n} , \quad \delta^j = \sum_n d_n^j \phi_{j,n} ,$$

where the coefficients of the two levels of resolution are related by

$$c_n^j = \sum_k h_{k-2n} c_k^{j-1} , \quad d_n^j = \sum_k g_{k-2n} c_k^{j-1} \qquad (2)$$

and $g_n = (-1)^n h_{-n+1}$. Obviously, h and g are the low-pass and high-pass filters respectively. The decimation by a factor of 2 corresponds to a down-sampling when going from one level to the next coarser one. This decomposition can be continued using the relation $V_j = V_{j+1} \oplus W_{j+1}$ and so on until a given level $J > j$, obtaining the following approximation for f:

$$f^{j-1} = \delta^j + \cdots + \delta^{J-1} + \delta^J + f^J$$

The inverse operation, the reconstruction of f^{j-1} from f^j and δ^j, is simply:

$$c_k^{j-1} = \sum_n (h_{k-2n} c_n^j + g_{k-2n} d_n^j) \qquad (3)$$

Filters for analysis and reconstruction satisfying the relations (2) and (3) are called quadrature mirror filters. Furthermore, they constitute a subband filtering scheme with exact reconstruction.

3.3   Symmetry and Exact Reconstruction

In practice one is mainly interested in reconstruction filters which have compact support. Orthonormal wavelet bases with compact support can be computed following algorithms given by Daubechies [6], where one also finds many examples. In her work she also shows that compact supported orthonormal wavelet bases cannot be symmetric. More precisely, they are not *linear-phase* filters [4]. For many applications, such as compression of Calderon-Zygmund and other pseudo-differential operators in numerical applications [1], asymmetric wavelets work very well. For other applications, such as digital image processing, quantization errors or difficult handling of boundary conditions due to asymmetry, may be unacceptable.

3.4   Biorthogonal Wavelets

A subband filtering scheme with exact reconstruction and symmetry is only possible if one employs different filters for decomposition and reconstruction. In wavelet multiresolution analysis this can be achieved if one uses two *dual* wavelet bases. Each basis is associated with a multiresolution analysis. For the decomposition and reconstruction of functions we will be working with two scaling functions and two mother wavelets: ϕ, ψ and $\tilde{\phi}$, $\tilde{\psi}$, and with the corresponding filters h, g and $\tilde{h}$, $\tilde{g}$. The subband filtering scheme for biorthogonal wavelets will use those four reconstruction filters. The decomposition equation (2) now takes the form

$$c_n^j = \sum_k h_{k-2n} c_k^{j-1} \, , \; d_n^j = \sum_k g_{k-2n} c_k^{j-1} \, , \tag{4}$$

and the reconstruction equation (3) becomes

$$c_k^{j-1} = \sum_n (\tilde{h}_{k-2n} c_n^j + \tilde{g}_{k-2n} d_n^j) \, . \tag{5}$$

3.5   Higher Dimensions

There are different possible ways of constructing multidimensional wavelet bases starting from a basis for the one dimensional case. One construction consist in the tensor product of one dimensional multiresolution analyses rather than considering the wavelet basis (see [6]). In this case, dilations on the basis control all indices simultaneously. The multiresolution ladder in $L^2(\Re^3)$ consist of subspaces defined by $\mathbf{V}_j = V_j \otimes V_j \otimes V_j$ and the scaling function is given by

$$\phi(x, y, z) = \phi(x)\phi(y)\phi(z)$$

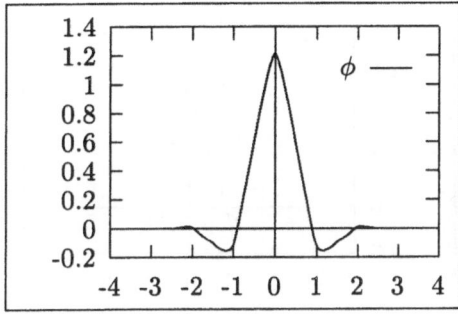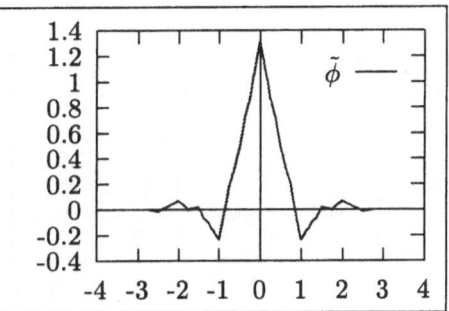

Figure 1: The scaling functions of the Daubechies spline case

The corresponding wavelet basis can be constructed considering all possible tensor products between the subspaces V_j and W_j. The three-dimensional *mother* wavelets are the following seven functions:

$$\psi_j^1(x, y, z) = \phi(x)\phi(y)\psi(z)$$
$$\psi_j^2(x, y, z) = \phi(x)\psi(y)\phi(z)$$
$$\vdots$$
$$\psi_j^7(x, y, z) = \psi(x)\psi(y)\psi(z)$$

4   The Volume Rendering Algorithm

In this section we demonstrate how wavelets can be used in the context of frequency domain volume rendering. We first consider the possibility of using scaling functions of biorthogonal wavelets as high quality reconstruction filters. We then present a method of improving the performance in the interpolation step based on the hierarchical wavelet decomposition of the volume data. The work in this section is motivated for the filtering properties of wavelets which are successfully used in the field of signal processing. Wavelets, or *small waves*, are functions which generate $L^2(\Re)$ and have the properties of localizing signals in space and frequency domain.

4.1   Wavelets as Reconstruction Filters

As we have seen in section 2 the algorithm for frequency domain volume rendering consists of the interpolation of a 2D slice P from the 3D spectrum F. Since we actually do not have the continuous F but its sampled version $\tilde{F}$, we need to reconstruct F from $\tilde{F}$ by convolution with a filter H. In section 3 we have shown that it is possible to construct dual scaling functions with compact support and linear phase, i.e. symmetric, by means of biorthogonal wavelets. Obviously, we are only interested in those pairs of functions were *leakage effects*

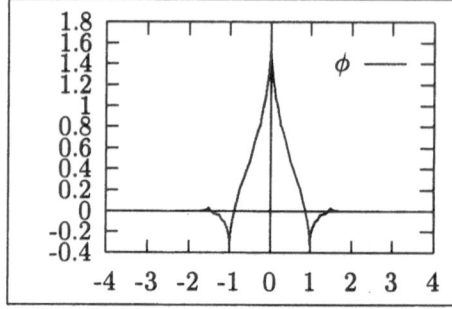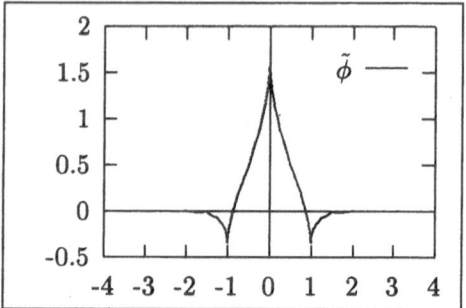

Figure 2: The scaling functions of the Burt family

are not relevant, that is, we look for filter functions which are as regular as possible but still have a relative small compact support. If ϕ is the scaling function, then the reconstructed F can be written as

$$F(\vec{x}) = \tilde{F} * \phi = \sum_{\vec{n}} F(\vec{n}) \phi_{0,\vec{n}}(\vec{x})$$

The scaling functions are being used as reconstruction filters, due to their linear phase and their *good* behavior in the space and frequency domain, there is no relation with multiresolution analysis, yet. For testing purposes we have implemented two families of biorthogonal wavelets. The first one was termed *a variation on the spline case* by Daubechies [6], where the scaling function ϕ, has a support width of 6 and $\tilde{\phi}$ a support width of 8 (see figure 1). The second class of functions is called the Burt family, which is defined with a support width of 4 for ϕ and 6 for $\tilde{\phi}$ (see figure 2). In order to see the behavior of the scaling functions of the Daubechies spline case in the frequency domain we show the Hartley transform their in figure 3.

4.2 A Hierarchical Algorithm

As we have mentioned in the introduction, frequency domain volume rendering algorithms spend most of their computing time interpolating the 2D slice from the 3D spectrum. The complexity of this process is $O(N^2)$, saying that there are N^2 samples to be interpolated. A speed-up of the algorithm can be achieved, if the number of samples to be interpolated can be reduced without a serious loss of image quality. To achieve this we apply a subband filtering scheme with two channels and decimation by factor of 2, also called down-sampling. This scheme is based on the wavelet transform, where we employ orthonornal and biorthogonal wavelet bases with compact support.

In the following the samples of f will be written as:

$$\mathbf{c}^0 = \left\{ c^0_{l,m,n} = f(l,m,n); l,m,n = 0, \cdots (N-1) \right\}.$$

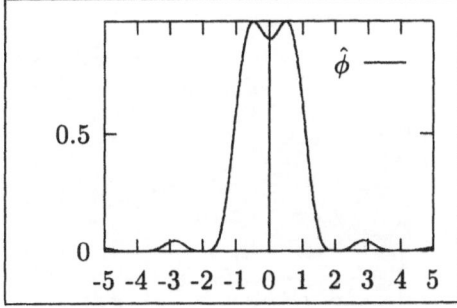 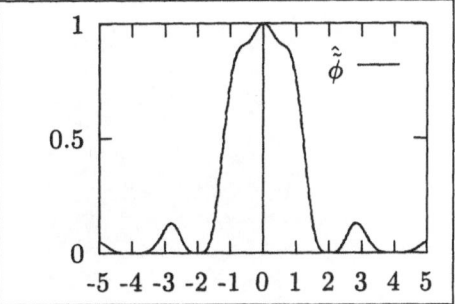

Figure 3: Hartley transform of the scaling functions of figure 1

Applying a wavelet decomposition using an orthonormal or a biorthogonal basis (4) the sequence of smooth coefficients

$$\mathbf{c}^1 = \{c^1_{l,m,n}; l, m, n = 0, \cdots, \frac{N}{2} - 1\}$$

and of detail components

$$\mathbf{d}^{\lambda,1} = \{d^{\lambda,1}_{l,m,n}; \lambda = 1, \cdots, 7; l, m, n = 0, \cdots, \frac{N}{2} - 1\}$$

are obtained. While the original data $\mathbf{c}^0$ consist of N^3 data samples, the eight transformed components $\mathbf{c}^1$ and $\mathbf{d}^{(1,\cdots,7),1}$ have $(\frac{N}{2})^3$ samples each.

Since the h coefficients in (4) act as a low pass filter the coefficients $\mathbf{c}^1$ contain all the information corresponding to the low frequencies of the spectrum of $\mathbf{c}^0$, but they are only $1/2^3$ of the number of the original coefficients. Therefore, we will base our volume rendering on this reduced data set. In the frequency domain the transform F^1 will also contain only $1/2^3$ samples of F. Thus, interpolating a 2D slice from F^1 will speed-up by a factor of 4. Due to the fact that a Hartley transformation pair satisfies the Rayleigh equation, and since the wavelets are actually a Riesz bases, we can estimate the quality of the approximation by:

$$\| F - F^1 \|^2 = \| \mathbf{c}^0 - \mathbf{c}^1 \|^2 \leq B \sum_{\lambda} \sum_{l,m,n} \| d^{\lambda,1}_{l,m,n} \|^2, B \geq 0.$$

It is a well known property of wavelets, that smooth data without large gradients result in small coefficients $d^{\lambda 1}_{l,m,n}$, which is used in wavelet data compression algorithms. The information contained in the $d^{\lambda 1}_{l,m,n}$ corresponds to the higher frequencies of the spectrum, which are lost in this case. Therefore, sharp edges and steep intensity gradients are smoothed out resulting in blurred images.

The decomposition step can be repeated a number of times obtaining at each level coarser representations of the input data. If one starts with a data set of dimension $256 \times 256 \times 256$, for example, after three steps one has a data set of dimension $32 \times 32 \times 32$. Rendering this data, however will produce images

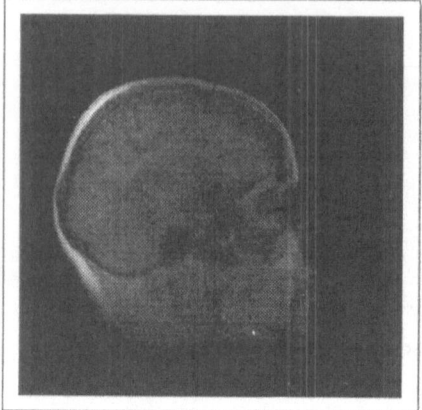 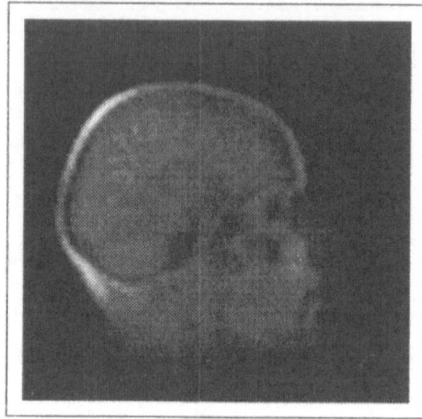

Figure 4: Images produced using the Burt wavelet for full data resolution (left) and after two decomposition steps (right)

with unacceptable quality. Using a thresholding technique, that means only saving those coefficients of the detail components which are above a user given threshold, it is possible to obtain a high data compression rate [12, 15, 13].

The important question of how to choose the wavelet basis can be answered by considering the wavelet decomposition as a filtering operation. The result of the decomposition depends on the properties of the basis functions used in the wavelet transform. If the basis function have a large number of vanishing moments, they will have good approximating properties which corresponds to a higher compression potential, but this function will lack regularity. More regularity in the basis functions corresponds to smooth elementary building blocks in the decomposition which introduce fewer quantization errors. If orthonormal Daubechies wavelets filters with a large number M of vanishing moments

$$\int \psi(x)x^l dx = 0 \, ; \, l = 0, \cdots, M$$

are used, a good data compression rate can be achieved [1]. The detail components will be very small or zero except for a few of them which correspond to regions of rapid data variation. The images produced using F^1 will have lost this information, which is visible as blurring. On the other hand, the biorthogonal wavelets of the Burt family or of the spline case present some regularity properties. Filtering with bases, which are more regular, but do not have a large number of vanishing moments, will therefore produce many relatively small detail components which cannot be neglected. This means, that more high frequency information is contained in the smooth component, leading to sharper images for the cost of reduced compression.

Concluding, the proposed method consists of the following steps:

- apply the wavelet transform to the input data a number of times

- perform data compression by thresholding

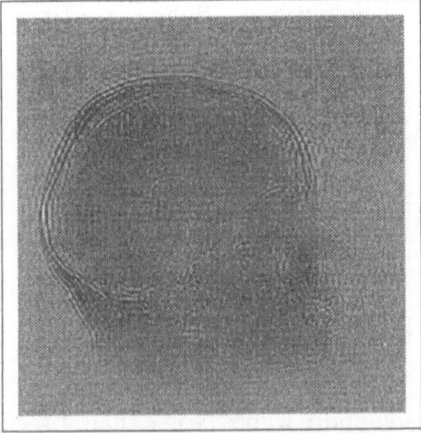 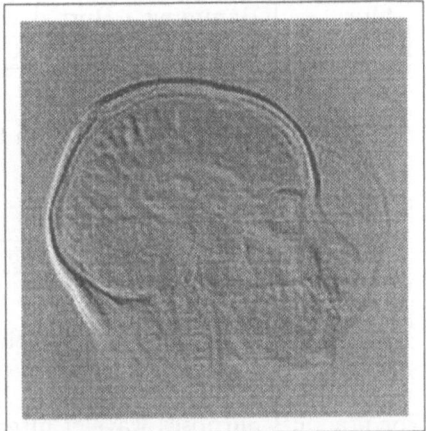

Figure 5: Difference between high resolution image and images generated from the first smooth component of the decomposition using Burt (left) and Daubichies (right) wavelets

- apply the frequency domain rendering algorithm to the smooth component, using wavelets as reconstruction filter

- if higher image quality is desired, reconstruct the data to a higher level of resolution using the fast wavelet transform, and repeat the rendering process

5 Results

In this paper we have described two ways of improving existing frequency domain volume rendering algorithms by using wavelets. Employing wavelets for the design of reconstruction filters reduces aliasing artifacts while maintaining efficiency. Additional speed-up results from a wavelet decomposition of the input volume. We have implemented the described methods using C++ and an object oriented approach on a SGI Indigo with a 100 MHz R4000 and investigated the properties of various wavelet bases.

With respect to the reconstruction process we tested three filters: the traditional Hamming window, and four scaling functions of biorthogonal wavelets, the scaling functions of the spline case of the Daubechies wavelets with support width 6 and 8, and the Burt family of support width 4 and 6 (see figures 1 and 2). Using the standard MRI data set of a human head containing $256 \times 256 \times 128$ samples and generating images of size 512×512 we found small differences in image quality. However, the wavelet filters show a flat-topped behavior, which introduces less quantization errors than the needle-shaped Hamming window, which requires many preprocessing operations such as data premultiplication.

filter	Hamming	Burt	Burt Dual	Spline case	Spline case Dual
time in s	10.12	21.48	61.73	61.84	136.63

Table 1: 2D slice interpolation time with respect to different filter sizes

data size	$256^2 \cdot 128$	$128^2 \cdot 64$	$64^2 \cdot 32$
time in s	21.48	5.87	1.95

Table 2: 2D slice interpolation time for different resolution levels

For practical purposes wavelet filters with a support width larger than 4 seem not to be worth the increased computational cost. In table 1 we give the increasing CPU time with respect to the filter size.

For implementing the hierarchical part of the algorithm we have used four wavelet families. The orthonormal Daubechies wavelets with 2 vanishing moments and support width 4, and with 3 vanishing moments and support width 6, the biorthogonal spline wavelets and the Burt family. The speed-up factor of 4 as predicted in section 4 for each decomposition step was confirmed (see table 2). The image quality of the first smooth component is quite acceptable, almost as good as the images obtained from the full resolution. In addition, differences between the results produced using the different wavelet bases can be observed. Biorthogonal bases with more regularity properties, such as the spline case using $\tilde{\phi}$ in the decomposition step, produce sharper images with a better resolution. The images derived from the second decomposition step (with a data reduction of a factor of 64) show strong blurring effects and should only be used for data exploration at interactive rates (see figure 4). The properties of the different wavelet bases with respect to data compression and smoothing can be seen in the difference images (figure 5). As expected, orthonormal Daubechies wavelets coefficients are almost zero everywhere except for regions with large gradients, introducing stronger blurring effects than the Burt wavelets, which display an overall better approximation to the high resolution image.

The authors wish to thank Takashi Totsuka for many helpful sugestions and comments concerning the implementation details of standard frequency domain volume rendering. Discussions with Günther Greiner on wavelets were very useful in different development stages of this project.

References

[1] Beylkin, G. and Coifman, R. and Rokhlin, V. Wavelets in Numerical Analysis. In: M. B. Ruskai, G. Beylkin, R. Coifman, I. Daubechies, S. Mallat, I. Meyer, L. Raphael (eds.), Wavelets and Their Application, pp. 181–210. Jones and Bartlett Publishers, Boston and London (1992).

[2] Bracewell, Ronald E. The Hartley Transform. Oxford University Press (1986).

[3] Brigham, E. Oran. The Fast Fourier Transform. Prentice-Hall Inc. (1974).

[4] Chui, Charles K. Wavelet Analysis and its Applications I: An Introduction to Wavelets. Academic Press, Inc. (1992).

[5] Civanlar, M.R. and Nobakht, R. A. Optimal Pulse Shape Design using Projections onto Convex Sets. In: ICASSP 88 Proceedings, vol. 3, pp. 1874–1877. IEEE ComputerSociety Press (1988).

[6] Daubechies, Ingrid. Ten Lectures on Wavelets. Society for Industrial and Applied Mathematics (1992).

[7] Dunne, Shane and Napel, Sandy and Rutt, Brian. Fast Projection of Volume Data. In: Proceedings of the First Conference on Visualization in Biomedical Computing, pp. 11–18. IEEE Computer Society, IEEE Computer Society Press (1990).

[8] Levoy, M. Display of Surfaces from Volume Data. Computer Graphics & Applications, 8(3), 29–37 (1988).

[9] Mallat, S. G. A Theory for Multiresolution Signal Decomposition: The Wavelet Representation. IEEE Transactions on Pattern Analysis and Machine Intelligence, 11(7), 674–693 (1989).

[10] Malzbender, T. Fourier-Volume-Rendering. ACM Transactions on Graphics, 12(3), 233–250 (1993).

[11] Marschner, S. R. and Lobb, R. J. An Evaluation of Reconstruction filters for Volume Rendering. In: R. D. Bergeron, A. E. Kaufman (eds.), Visualization '94, pp. 100–107. IEEE Computer Society, IEEE Computer Society Press (1994).

[12] Muraki, S. Volume Data and Wavelet Transforms. IEEE Computer Graphics and Applications, 13(4), 50–56 (1993).

[13] Tao, H., Moorhead, R. T. Progressive Transmision of Scientific Data Using Biorthogonal Wavelet Transform. In: A. Kaufman, W. Krüger (eds.), 1994 Symposium on Volume Visualization, pp. 93–99. ACM SIGGRAPH (1994).

[14] Totsuka, T., Levoy, M. Frequency Domain Volume Rendering. Computer Graphics, 27(4), 271–78 (1993).

[15] Westermann, R. A Multiresolution Framework for Volume Rendering. In: A. Kaufman, W. Krüger (eds.), 1994 Symposium on Volume Visualization, pp. 51–58. ACM SIGGRAPH (1994).

[16] Westover, L. Footprint Evaluation for Volume Rendering. Computer Graphics, 24(4), 367–376 (1990).

Order of Pixel Traversal and Parallel Volume Ray-tracing on the Distributed Shared Volume Buffer

Hansong Zhang, Shenquan Liu *

CAD Lab., Institute of Computing Technology

Academia Sinica, P. O. Box 2704, Beijing 100080, P. R. China

zhs@osf.cnc.ac.cn

Abstract. The distributed shared volume buffer (DSVB) is a software package we developed to facilitate general, parallel volume ray-tracing on networked workstations. It is internally implemented with message-passing and adopts the cache-coherent shared memory model. Thus the cache efficiency of volume data access is of utter importance to the performance of a DVSB-based ray-tracer. For a given data set, the data access behavior of a volume ray-tracer depends mostly on the way in which pixels of the image are traversed. This paper addresses the cache coherence problem and compares three kinds of pixel traversal order: one-way, two-way and along a space filling curve. Experiments show that traversing pixels along a space filling curve (e.g. a Hilbert curve) greatly enhances cache efficiency especially when size of the cache is small compared to that of the volume data, and in the meantime greatly simplifies task distribution and management.

Keywords: Pixel Traversal, Parallel Volume Ray-tracing

1 Introduction

Raytracing volume data can produce impressive images that help a lot in understanding the data [10, 11]. The cost of ray-tracing, however, is quite expensive both in time and in memory space. A high-quality image may take as much as hours to render. Volume data sets are typically large and get easily too large to hold in the main memory of an average workstation. To solve these problems, data distributed and parallel volume ray-tracing algorithm have been developed for modern parallel computers. A network of workstations, which is logically equal to an MIMD, distributed memory, message-passing based multicomputer, provides a cost-effective and powerful environment for parallel volume rendeing.

The great potential of multicomputers and networked workstations must be fully tapped by proper parallel applications. There are plenty of existing serial

---

*This work was supported by National Natural Science Foundation of China.

volume ray-tracing programs. Unfortunately, however, these serial ray-tracers cannot be easily ported to a distributed-memory parallel environment when the data are distributed among the processing nodes. The direct access of volume data in a continuous, flat address space, which is implicitly assumed in any serial ray-tracer, is not readily available in such an environment. Replicating the whole data in every node's local memory will elliminate the problem, but due to the limited size of the local memory this is not a desireable approach. So the major question for data distributed parallel ray-tracers is: what should I do if the data needed is not in the local memory? One solution is to distribute task in such a way that computation on a node need only local data [1, 3, 4, 5]. Or computation itself can be sent to the node where the requrired data reside [7]. All these approaches lead to highly specific algorithms that bear little resemblence to a serial algorithm. Besides, in general volume ray-tracing (in contrast to simple ray-casting), light behaviors like reflection, refraction and shadowing are taken into account. The ray through any pixel may access any data. In this case it is impossible to construct tasks that access only local data. Porting a serial ray-tracer to a shared memory multicomputer (like DASH)[2, 9], on the other hand, is direct and simply. Aside from task distribution, and the parallel ray-tracer is much the same as a serial one.

The distributed shared volume buffer (DSVB) is a software package we developed to provide a virtual shared memory of volume data for networked workstations. The volume data is physically distributed among machines (nodes) that participate in ray-tracing, but through the DSVB interface, the ray-tracer on each node can regard all data as local. DVSB incorporates the cache-coherent shared memory model [2]. Authors of [6] implemented a virtual memory based on software cache on Fujitu AF100, though in their case the nessesity of global data access arises from dynamic load balancing, but not from global illumination. Section 2 gives a breif description of DVSB.

A software cache is a buffer in the local memory for recently used data that are not local. If the requested non-local data are found in the cache, then a remote data fetching is spared. The efficiency of cache is crucial to the performance of DVSB, and to applications based on it. This is particularly true when the parallel environment is a group of workstations connected by moderate-performance network (e.g. 10Mb/s Ethernet), where communication cost is high. Besides the design of the cache itself, the way how volume data is accessed greatly affects cache efficiency. In a volume ray-tracer, the data access pattern is mostly decided by the order in which pixels are ray-traced. Section 3 and section 5 discuss the pixel traversal order problem and compare three kinds of order.

2 The Distributed Shared Volume Buffer

The structure of our cache-coherent shared volume buffer is shown in figure 1. The buffer is divided into "pages"; page is the unit of network transfer of volume data. The volume data does not lay linearly in the virtual buffer. Instead, it is stored in terms of blocks, with one block residing in one page. Each block contains a $K \times K \times K$ sub-set of data, where K is adjustable and is used to decide the size of a page. The original volume data is padded with zeros to make its X, Y, and Z dimensions multiples of K. Dividing the data into 3D blocks takes into account the coherence of data access by a ray traveling through the volume. In linear division of volume data most pages contain 2-D piece of data, among which very few can be used by a ray.

Applications based on DSVB directly request data values, and don't have to know about the internal organization of data. The most commonly used interface to DSVB is

$$\texttt{VolVal(P, V)}$$

where P is a position vector in 3D space and V is an array of 8 integers. On return, the function stores in V the eight values on the corners of the volume cell that surrounds P.

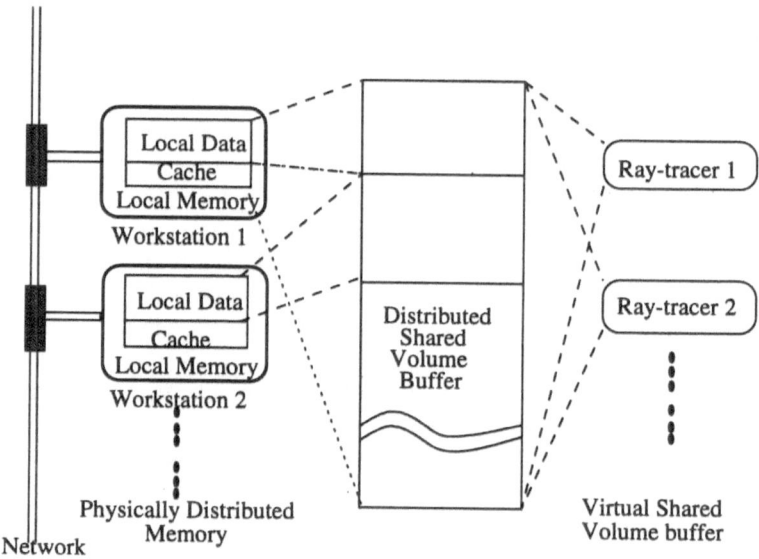

Figure 1: The distributed shared volume buffer

We call each of the networked workstations that take part in ray-tracing a node. When DSVB is initialized, every node gets a group of adjacent blocks of volume data. The data are distributed evenly and without special strategy. During computation, when on a node (say node A) the requested data is not in the local memory, DSVB looks for it in the cache. If it is in the cache (this is called a cache hit), then DSVB will return the cached value. If not, the the data have to be fetched from a remote node, and a situation similiar to a page fault in an operating system occurs. DSVB first finds out where the data reside as local (simple calculation, since the data blocks are distributed evenly), and send a page request to the node (say node B). Node B then send back the page requested by A. Node A, on getting the page, tries to allocate an item in the cache for it. If the cache no longer has free items, an LRU (Least Recently Used) algorithm is used to discard the content of an old item to make space. After the page is cached, DSVB return the requested data values, thus ending the data retrieving. Since a remote data request reqires a round trip on the network, it consumes considerable time. So cache efficiency (marked by the ratio of cache hits to total non-local data accesses, or amount of data tranferred because of cache misses) is critical for DSVB's performance.

3 Order of Pixel Traversal

Most serial ray-tracers pay little attention to the order of pixel traversal. They traverse the pixels on the image plane according to a "natural" order, e.g. row by row. For a DSVB based parallel volume ray-tracer, however, the order deserves a closer look, because it has great impact on the ray-tracer's data access coherence. High coherence means that recently accessed data, which are in the cache, are likely to be used again, thus reducing the times of remote data fetching through the network.

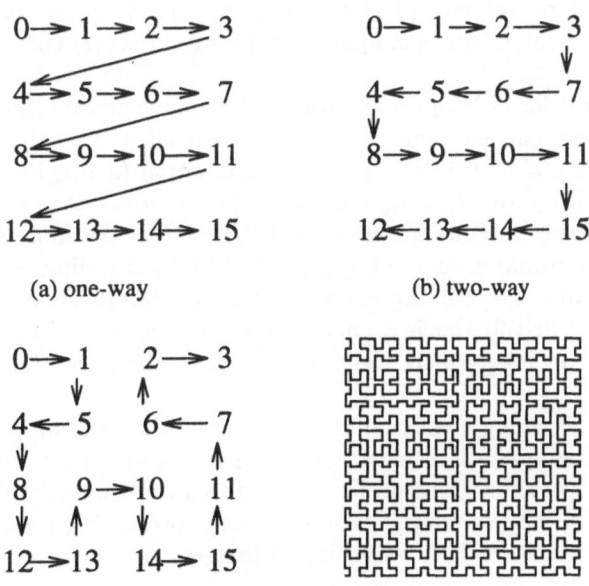

Figure 2: Pixel traversal order

As a matter of fact, rays through pixels that are close to each other access similiar data, or in other words, have high access coherence. So, the coherence of m pixels $\{P_1, ..., P_m\}$ can be measured by the weighted average distance between these pixels:

$$C = \sum p_{ij}|P_iP_j|, 1 \leq i < j \leq m \qquad (1)$$

where $| P_iP_j |$ is the distance between P_i and P_j, $p_{i,j} \geq 0$, $\sum p_{ij} = 1$, $1 \leq i < j \leq m$. The coherence of a pixel traversal path for $n(n \geq m)$ pixels can be calculated by sliding an measure interval of m pixels along the path, and calculate C for each group of adjacent m pixels. This lead to a coherence function $C(s, m), s \in \{1, ..., n-m+1\}$, with s indicating the starting point of the measure interval. A good traversal path's $C(s, m)$ should have low flunctuation and small average.

Here we examine three kinds of pixel traversal order (see figure 2). The first is the "natural" order which goes through the pixels row by row, and we term it *one-way* in order to distinguish it form the second type of order. The second type also travels row by row, but takes inverse directions on adjacent rows. This we call the *two-way* order. The third way of traversal is to follow a space filling curve.

Let $I = [0, 1]$, a space filling curve is a continuous function $c(t)$ with domain I and range $I \times I$. When the mapping between I and $I \times I$ is 1-1, then the curve is simple. A simple space filling curve establishes a one to one correspondence between a point on interval $[0, 1]$ and a 2-D point in square $[0, 1] \times [0, 1]$. So, when t changes from 0 to 1, a simple space filling curve $c(t)$ visits uniquely every point in $I \times I$.

Here we consider only approximations of a simple space filling curve defined above. Often we call such an approximation itself a space filling curve. Let $I_n = \{\frac{1}{n}, \frac{1}{n-1}, ..1, n \in N\}$, an approxiamtion curve (a polyline) defines a 1-1 maps between $I_{n \times n}$ and $I_n \times I_n$. Let $N_n = \{1, ..., n, n \in N\}$, then after simple transformation the polyline maps $N_{n \times n}$ to $N_n \times N_n$. An image's domain is in fact the set of coodinates denoted by $N_n \times N_n$. So by traveling along the polyline approximation of a space filling curve, we traverse the pixels in an ordered and unique way. Intuitively, such a curve gives good coherence for any cluster of adjacent pixels on its path by the standard of (1) because of its self-entwined structure.

Visiting pixels along a space filling curve has been used in image processing [12], where typical operations occur on subregions of an image. By using a space filling curve, a 2-D subregion can be mapped to an 1-D interval. This reduction of dimensionality lead to simplification of image processing methods. [13] used space filling curves in digital halftoning of images.

4 Task Allocation

Our ray-tracer uses distributed task pools and task stealing paradigm for task allocation and dynamic load balancing. On start-up, every node obtains approximately the same number of pixels, which comprises the local task pool. If the local task pool of node K is exhausted, it asks node (K+1) *mod* n (n is the number of nodes) for more work. If this request fails, then node (K+2) *mod* n is asked, an so on. This results in a "food chain" among nodes. In this way computing load is balanced.

Obviously, it would lead to very poor data access coherence if we visit the entire image in the one-way or two-way order. For these two kinds of traversal, we first divide the image plane into small sub-blocks (e.g. 16×16 ones) of pixels, and use block as the unit of task allocation. The blocks are evenly distributed to nodes on start-up. The pixels in a block are visited in one-way or two-way order.

For tranversing along a Hilbert curve, task management is dramatically simpler: the division and distribution of an image are reduced to operations on an 1-D interval ($[1, n \times n]$ for an $n \times n$ image); a node's local task pool is as simple as two integers indicating the starting and ending number.

5 Results

We tested the three orders of pixel traversal on two scenes. One scene is a $256 \times 256 \times 109$ head surround by 4 mirrors (see figure 5). Another is a $128 \times 128 \times 197$ data set of animal trunk with reflections in 3 mirrors (figure 6). In the tests

cache size / data size	Data Transfer (Mbyte)			Rendering Time(sec.)
	one-way	two-way	Hilbert	Hilbert
%5	-	-	1108	6262
%10	281.6	240.1	132.7	226
%15	207.4	175.3	88.1	194
%20	159.4	145.9	73.1	171
%25	131.6	120.7	60.3	154
%30	126.7	113.8	56.9	150
%35	119.3	108.7	50.6	145
%40	113.9	96.5	47.3	145
%50	102.7	85.9	40.4	142
%60	80.5	75.3	36.1	138
%70	71.8	68.2	34.3	137
%80	61.4	60.0	34.3	132
%90	56.1	56.2	34.1	128

Table 1: Test results with the head data

we used 8 SGI Indigo workstations (with 50MHz R4000 CPUs) networked by 100Mb/s Ethernet. Test results are listed in table 1 and table 2, and illustrated in figure 3 and figure 4.

The impact of different orders on cache efficiency is shown by the amount of volume data transfer that is incurred by cache misses. Rendering time (excluding

cache size / data size	Data Transfer (Mbyte)			Rendering Time(sec.)
	one-way	two-way	Hilbert	Hilbert
%5	-	-	-	-
%10	-	-	-	-
%15	541.3	484.0	241.6	298
%20	214.4	185.8	74.4	127
%25	175.9	147.3	55.8	112
%30	139.7	120.0	48.0	107
%35	110.4	80.2	40.4	110
%40	86.9	64.5	35.6	99
%50	57.6	48.6	30.4	99
%60	43.1	39.1	27.9	93
%70	36.1	32.9	25.4	92
%80	30.8	28.5	23.8	88
%90	24.3	24.1	22.6	87

Table 2: Test results with the trunk data

time for data distribution) for Hilbert curve traversal is also given. The tests are performed on ordinary work days when all machines and the network are busy, whose load may vary considerably in the process of our tests. The serial version of our renderer, which directly accesses a volume data array, took 753 and 380 seconds to render the head and the trunk scene, respectively.

In these tests all other parameters are fixed except for the cache size. We use $32 \times 32 \times 32$ blocks, so the size of a page is 32K bytes. This is the amount of data transferred in a single remote data fetching. Under this block size, the head data is padded to $256 \times 256 \times 128$ ($8 \times 8 \times 4$ blocks), and the trunk to $128 \times 128 \times 224$ ($4 \times 4 \times 7$ blocks). Image block size (for one-way and two-way traversal) is 16×16. The image size is 512×512.

Results show that the two-way traversal is a bit more efficient than one-way, and Hilbert curve traversal gives significant performance boosts especially when size of the cache is small compared to the volume data. With Hilbert traversal and small cache size, data traffic drops steeply as the cache becomes larger, showing high cache efficiency. Our tests made use of 8 machines, and the volume data are evenly distributed among them, so when the cache size is larger than $\frac{7}{8} = 87.5\%$ of the data size, each node can accomodate all the volume data. In this case, the maximum amount is $M = (\frac{n-1}{n}D)n = (n-1)D$, where D is the size of the volume data and n the number of nodes. This occurs when all nodes accessed all pages. For the head data, $M \approx 56$Megabytes. From the test results we can see that the Hilbert order result in lower data transfer than M when size of the cache if greater than 30% that of the head data. This implies that the pixels traced by a node are well "clustered", so that quite some non-local data are never accessed and thus never fetched. For the trunk tests, however, the

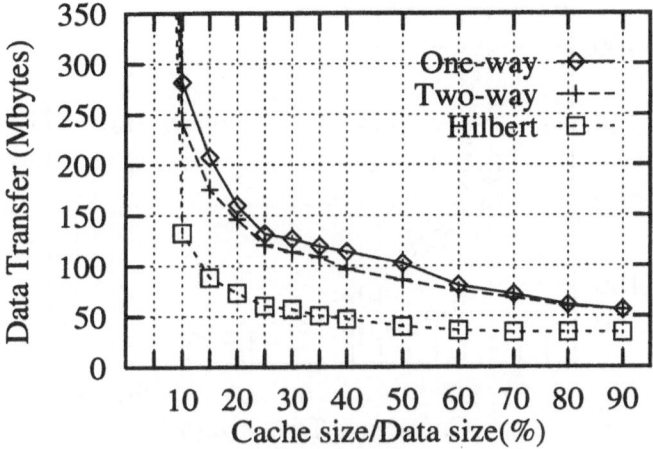

Figure 3: Amount of data transferred in tests with the head data.

data traffic at the high end of cache size, is not much less than M ($\approx$ 24.5Mb). This is because the setting of the trunk scene incurs more complicated reflection phenomenon than the head scene. In other words, the data are accessed more globally because of the reflections. Note that though the trunk data is smaller than the head, the data traffic in its tests is no less, sometimes even greater than tests with the head.

When the cache is too small, DVSB's performance drops dramatically because of *trashing*. Trashing means that the cache has to discard pages that are soon to be used again, thus giving rise to repeated remote data fetching. The missing data from table 1 and table 2 (marked by "-") indicate that trashing occured, and we have to abandon the tests either because the system manager and other users complained about network jams or we ourselves could not wait any longer.

6   Conclusion

In this paper we introduce the distributed shared volume buffer (DSVB) we developed to facilitate general parallel volume rendering. We discuss the relationship between the order in which a ray-tracer visits pixels and the cache efficiency of DSVB. Three types of traversal are tested: one-way, two-way and space filling curve. Experiments show that traversing along a space filling curve greatly enhances cache efficiency, and thus the performance of the ray-tracer.

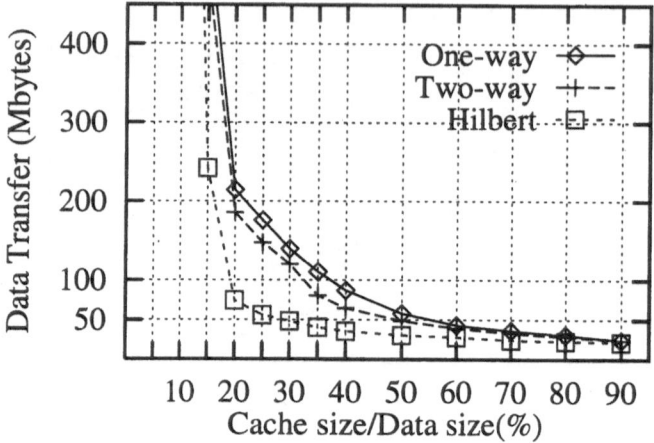

Figure 4: Amount of data transferred in tests with the trunk data.

References

[1] K.-L. Ma, J. S. Painter, C. D. Hansen and M. F. Krogh, Parallel Volume Rendering Using Binary-Swap Compositing, IEEE Computer Graphics and Applications, July 1994, pp. 59-68

[2] J. P. Singh, A. Gupta and M. Levoy, Parallel Visualization Algorithms: Performance and Architectural Implications, IEEE Computer, July 1994, pp. 59-55

[3] U. Neumann, Parallel Volume-Rendering Algorithm Performance on Mesh-Connected Multicomputers, Proc. Parallel Rendering Symp., ACM, New York, 1993, pp. 97-104

[4] W. M. Hsu, Segmented Ray Casting for Data Parallel Volume Rendering, Proc. Parallel Rendering Symp., ACM, New York, 1993, pp. 7-14

[5] E. Camahort and Chakravarty, Integrating Volume Data Analysis and Rendering on Distributed Memory Architectures, Proc. Parallel Rendering Symp., ACM, New York, 1993, pp. 23-26

[6] B. Corrie and P. Mackerras, Parallel Volume Rendering and Data Coherence, Proc. Parallel Rendering Symp., ACM, New York, 1993, pp23-26

[7] D. Badouel, K. Bouatouch and T. Priol, Distributing Data and Control for Ray Tracing in Parallel, IEEE Computer, July 1994, pp. 59-55

[8] G. Vézina, P. A. Fletcher and P. K. Robertson, Volume Rendering on the MasPar MP-1, 1992 Workshop on Volume Visualization, Boston, October 1992, pp. 3-8

[9] J. Nieh and M. Levoy, Volume Rendering on Scalable Shared-Memory MIMD Architectures, 1992 Workshop on V.lume Visualization, Boston, October 1992, pp. 9-16

[10] M. Levoy, Efficient Ray Tracing of Volume Data, ACM Transactions on Graphics, July 1990, pp. 245-261

[11] R. A. Drebin, L. Carpenter and P. Hanrahan, Volume Rendering, Computer Graphics, vol. 22, August 1988, pp. 65-74

[12] R. J. Stevens, R.A.Lehar, F. H. Perston: Manipulation and Preservation of Multidimensional Image Data Using the Peano Scan, IEEE Transaction on Pattern Analysis and Machine Intelligence, May 1983, pp. 520-526

[13] L. Velho, J. Gomes, Digital Halftoning With Space Filling Curves, Computer Graphics, Vol. 25(4), July 1991, pp. 81-90

Editors' Note: see Appendix, p. 156 for coloured figures of this paper

Visualization of local stability of dynamical systems

Georg Fischel, Eduard Gröller

Institute of Computer Graphics
Technical University Vienna
Karlsplatz 13/186/2, A-1040 Vienna, Austria

Abstract. Several methods for visualizing local stability properties of dynamical systems are presented. The calculation of characteristic values of local stability for linear and nonlinear systems is discussed. Two principles of visualizing local stability are introduced. The first principle is to display the estimated stability values directly by using scaled spheres or vectors. The second principle uses numerical analysis which generates portions of sweeps, that are deformed in dependence of local stability properties.

1. Introduction

Any system whose temporal evolution from some initial state is dictated by a set of rules is called a dynamical system. Dynamical systems can be found nearly everywhere in physics and other sciences: flow dynamics in fluids or gases, a simple pendulum or even human population numbers are dynamical systems. Dynamical systems can be classified according to several properties. An important property is the stability criterion. For most dynamical systems, the classification of stability has to take account that stability properties can be locally different within the system. There may be regions with a higher or lower degree of stability.

Visualizing this local degree of stability may give information whether a system is easily predictable in some regions or not and helps to understand the behavior of the whole system. In the next sections we will discuss how to estimate and to visualize stability properties of dynamical systems.

1.1 Classes of dynamical systems

Dynamical systems can generally be split into two classes. One can distinguish between continuous and discrete dynamical systems. Continuous systems can be described mathematically by a set of differential equations $\dot{x} = f(x)$. The solution of a system is then also called a flow because it is continuous in time.

Discrete dynamical systems are described by a set of difference equations $x_{n+1} = f(x_n)$. In this case the states are seperated in time by finite distances or intervals. One prominent example for a discrete dynamical system is $x_{n+1} = \mu x_n(1 - x_n), 0 \le \mu \le 4$, which is called the logistic equation [Peit92].

Another important classification is, whether a system is dissipative or conservative. Conservative systems are energy preserving like an undamped pendulum (Fig. 1 left). A dissipative system is energy consuming. The energy loss might for example be caused by friction which is the case for a damped pendulum (Fig. 1 right).

1.2 Phase space and trajectory

Phase space is used for the description of the behavior of a system in dependence of time t. Phase space is defined by assigning each variable of a dynamical system to a coordinate axis. The dimension of phase space is equal to the number of variables of the system. In the example of the pendulum (Fig. 1) the current position of the pendulum $x(t)$ is assigned to the horizontal axis and velocity $v(t)$ is assigned to the vertical axis. Thus the pendulum has a two-dimensional phase space.

The path of a point in phase space describes the temporal evolution of an initial state and is called trajectory in case of a continuous dynamical system and is called orbit in case of a discrete dynamical system.

When solving a dynamical system mathematically the solution is a specific trajectory or orbit which depends on the initial conditions of the system at time $t = 0$.

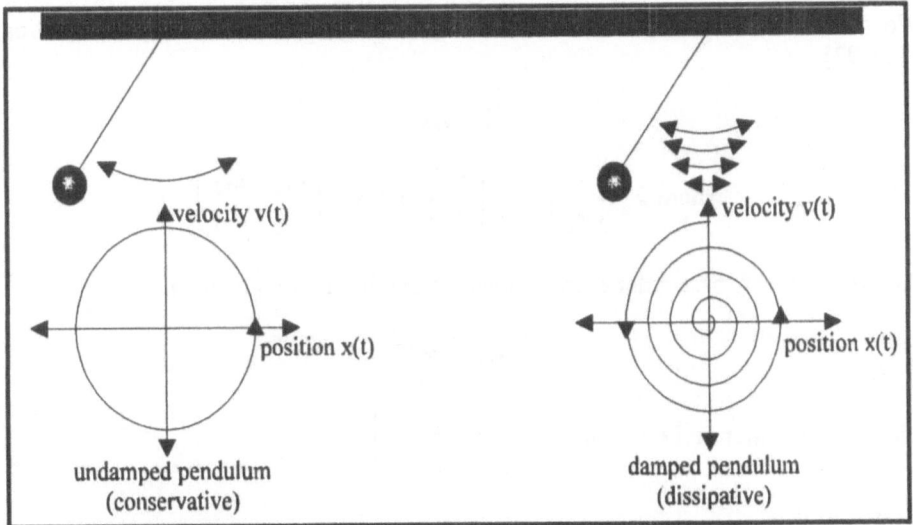

Fig. 1. Phase space and trajectory of a pendulum [Chao89].

1.3 Local stability and Lyapunov exponents

For dynamical systems described by linear homogenous first-order differential equations stability means, that a system reaches an equilibrium state independently from its initial condition. But more complex systems may have locally different stability properties. A system is locally stable/unstable if neighbouring trajectories converge/diverge. The local stability properties can be estimated by calculating the local Lyapunov exponents, which will be introduced in the next section.

1.4 Equilibrium states

A continuous dynamical system has an equilibrium state if the system of differential equations $\dot{x} = f(x)$ has a solution $f(x) = 0$. In this case the length of the tangent vector is zero and the system is in a so called equilibrium state (fixed point). Systems may have one or more equilibrium states. For homogenous systems the trivial solution $x = 0$ of $f(x) = 0$ is always an equilibrium state.

2. Determining stability

The first part of this section will describe (by using a 2D example) how to solve linear homogenous first-order differential equations. In the second part we will discuss how to estimate the stability of these systems. Finally we will deal with the calculation of the local stability of more complex nonlinear systems.

2.1 Solving linear first-order differential equations

We start with a linear, first-order homogenous system of differential equations [Tson92]

$$\dot{x} = Ax$$

$$\text{where } \dot{x} = \begin{pmatrix} \dot{x}_1 \\ \dot{x}_2 \end{pmatrix}, \ x = \begin{pmatrix} x_1 \\ x_2 \end{pmatrix} \text{ and } A = \begin{pmatrix} a_{11} & a_{12} \\ a_{21} & a_{22} \end{pmatrix}$$

For such a linear, first-order homogenous system the solution is given as

$$x(t) = ce^{\lambda t}$$

where λ is a scalar and c is a non-zero vector $\begin{pmatrix} c_1 \\ c_2 \end{pmatrix}$

It follows, that

$$\dot{x}(t) = \lambda ce^{\lambda t}$$

When substituted into $\dot{x} = Ax$ one gets

$$Ac = \lambda c$$

To calculate λ, the equation

$$|\mathbf{A} - \lambda\mathbf{I}| = 0$$

has to be solved. In other words, the required λ is the eigenvalue of $\mathbf{A}$ and the vector $\mathbf{c}$ is the corresponding eigenvector. In the above 2-dimensional example two eigenvalues, λ_1 and λ_2 result. If λ_1 and λ_2 are different, the corresponding eigenvectors are linearly independent [Lips77]. Both, $\mathbf{c}_1 e^{\lambda_1 t}$ and $\mathbf{c}_2 e^{\lambda_2 t}$, are solutions of $\dot{\mathbf{x}}(t) = \mathbf{A}\mathbf{x}$. Also any linear combination of the above solution is a solution of the linear system itself. Thus, the general solution can be written as

$$\mathbf{x}(t) = a_1 e^{\lambda_1 t}\mathbf{c}_1 + a_2 e^{\lambda_2 t}\mathbf{c}_2$$

If the system has dimension n the equation $|\mathbf{A} - \lambda\mathbf{I}| = 0$ is an equation of degree n. Thus, it has n solutions of λ. So the general solution of a homogenous, linear first-order system is

$$\mathbf{x}(t) = \sum_{i=1}^{n} a_i e^{\lambda_i t}\mathbf{c}_i$$

2.2 Possible values of λ_1, λ_2 and their meaning

In the above 2D example of a linear dynamical system the eigenvalues λ_1, λ_2 can be either both real or both complex. If they are both real and both are negative, then the final state $\mathbf{x}(t)$ approaches 0 as t goes to infinity independently of the initial condition $\mathbf{x}(0)$. The system is attracted to its equilibrium state (see Fig. 2). The equilibrium state is said to be asymptotically stable and is called a fixed point or node. If λ_1 and λ_2 are positive, then the final state $\mathbf{x}(t)$ approaches infinity as time goes to infinity. Independently from the initial condition the system will diverge and the equilibrium state will never be reached. Therefore it is not stable and the equilibrium state is said to be a repeller, i.e., any point close to the equilibruim state is repelled from it. If $\lambda_1 < 0 < \lambda_2$, then λ_1 tries to make the final states reach the equilibrium state, whereas λ_2 tries to repell the final states from the equilibrium state. This leads to evolutions that move towards the equilibrium state for a while and then move away. Such an unstable equilibrium state is called a saddle or hyperbolic point.

If λ_1 and λ_2 are complex numbers (i.e., $\lambda_{1,2} = \alpha \pm i\beta$) then $\mathbf{x}(t) = e^{\alpha t}(\mathbf{k}_1 \cos\beta t + \mathbf{k}_2 \sin\beta t)$ where $\mathbf{k}_1$ and $\mathbf{k}_2$ are appropriate vectors. Complex values indicate a rotation around the equilibrium state. The value of α indicates whether the points are moving away from or towards to the equilibrium state [Tson92].

Note, that for real λ_i the corresponding eigenvectors point into the direction of convergence or divergence.

110

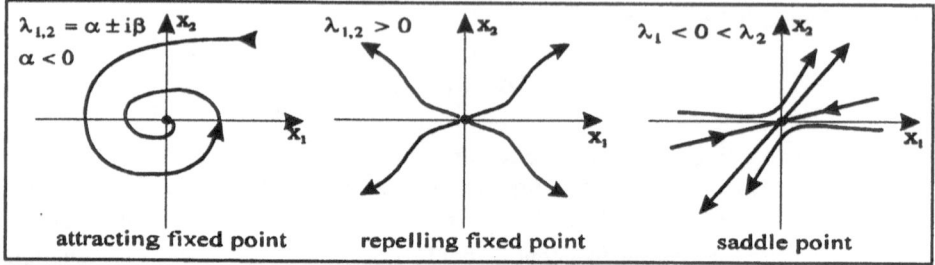

Fig. 2. Some equilibrium states and their stability [Tson92]: an attracting fixed point or node, a repelling fixed point and a saddle point.

2.3 Stability for nonlinear systems

In linear systems all λ_i remain constant for the whole system. Thus, the stability is the same everywhere within the system. When a linear system is stable, it reaches an equilibrium state independently from its initial condition. But most systems are nonlinear. Nonlinear systems have different stability at different locations within the system.

In general there is no analytic solution for systems of nonlinear differential equations. To estimate the local stability the system has to be linearized locally. This can be achieved by using the Jacobian matrix, which is made up of the partial derivatives of the differential equations. Given a system of differential equations

$$\dot{x}_1 = f_1(x_1, x_2, ..., x_n)$$
$$\dot{x}_2 = f_2(x_1, x_2, ..., x_n)$$
$$...$$
$$\dot{x}_n = f_n(x_1, x_2, ..., x_n)$$

then the corresponding Jacobian matrix is

$$J = \begin{pmatrix} \dfrac{\partial f_1}{\partial x_1} & \dfrac{\partial f_1}{\partial x_2} & \cdots & \dfrac{\partial f_1}{\partial x_n} \\ \dfrac{\partial f_2}{\partial x_1} & \dfrac{\partial f_2}{\partial x_2} & \cdots & \dfrac{\partial f_2}{\partial x_n} \\ \cdots & \cdots & \cdots & \cdots \\ \dfrac{\partial f_n}{\partial x_1} & \dfrac{\partial f_n}{\partial x_2} & \cdots & \dfrac{\partial f_n}{\partial x_n} \end{pmatrix}$$

The local solution of the linearized system is now

$$\Delta\dot{x} = J\Delta x$$

where $\Delta \dot{x} = \begin{pmatrix} \Delta \dot{x}_1 \\ \dots \\ \Delta \dot{x}_n \end{pmatrix}$ and $\Delta x = \begin{pmatrix} \Delta x_1 \\ \dots \\ \Delta x_n \end{pmatrix}$ is a perturbation relative to a point on a trajectory.

After linearization, the system is a homogenous, linear, first order-system so that for the calculation of stability the equation

$$J c = \lambda c$$

has to be solved (see section 2.1).

The λ_i are the eigenvalues of the Jacobian matrix and are called the local Lyapunov exponents [Tson92]. The Lyapunov exponents allow to estimate the local stability of a nonlinear system.

3. Visualization techniques

This section describes techniques of visualizing stability properties. For visualization the Rössler system was used, which is a chaotic dynamical system [Peit92].

3.1. Previous work

Several techniques for visualizing stability have been impelemented. In [Tson92] the trajectories of a Lorenz system are color-coded as follows: green represents the largest negative and red the largest positive Lyapunov exponent. In [Leeu93] a visualization technique is introduced, that displays several important local properties: velocity, rotation, acceleration shear and convergence. These properties are calculated by decomposing the Jacobian matrix. In [Schr91] the stream polygon is introduced. The stream polygon is an n-sided polygon with radius r. This polygon is deformed using a transformation matrix which is calculated using the Jacobian matrix. One important use of stream polygons is to sweep them along a trajectory to get the streamtube. Applying several transformations to the stream polygon gives a more or less deformed sweep. The kind of deformation, strain, shear strain, rotation, or a combination thereof gives information about the local behavior of a system. Further techniques on visualizing chaotic dynamical systems are described in [Fisc94] and [Groe94].

3.2. The Rössler system

The Rössler system (see plate 1) was introduced by Otto E. Rössler in 1976. It is only a fictive model, which is used to describe the properties of chaos in a simple continuous dynamical system [Peit92]. The system of differential equations for the Rössler system is

$$\dot{x} = -y - z$$
$$\dot{y} = x + ay$$
$$\dot{z} = b + xz - cz$$

where the parameters a, b and c are $a=0.2$, $b=0.2$, $c=5.6$. The corresponding Jacobian matrix for the Rössler system is

$$J = \begin{pmatrix} 0 & -1 & -1 \\ 1 & a & 0 \\ z & 0 & x-c \end{pmatrix}$$

It can be seen immideatly by looking at the Jacobian matrix that the eigenvalues only depend on the x- and z-coordinates, but not on the y-coordinate.

The equation $|J - \lambda I| = 0$ for the Rössler system, which is a 3-dimensional system, is an equation of degree 3. Therefore there exist three solutions for λ_1, λ_2 and λ_3. Either all three λ_i are real or λ_1 is real and λ_2, λ_3 are conjugated complex numbers.

If all three eigenvalues are real, then the local dynamics are as follows: neighbouring points of a trajectory converge/diverge in direction of the three eigenvectors which correspond to the λ_i. In case of one real and two complex λ_i, we have convergence/divergence in direction of the eigenvector corresponding to the real λ_1. The complex eigenvalues λ_2 and λ_3 induce a rotation with the eigenvector corresponding to λ_1 as rotation axis.

3.3. Visualization techniques

Two principles for visualizing local stability properties were implemented. The first one is the visualization of eigenvalues and eigenvectors of the Jacobian matrix, which denote direction and magnitude of convergence/divergence for a specific point in the system. The second principle is to observe and visualize the local evolution of points on neighbouring trajectories.

Visualizing Lyapunov exponents. The following method is used to display the magnitude of the biggest positive or negative Lyapunov exponents or the magnitude of their imaginary parts. First we choose, which exponents shall be displayed (i.e., the positive or negative exponents or their imaginary parts in case they are complex numbers). Then several points along a trajectory of the system are displayed as spheres. The radii of the spheres are proportional to the maximum magnitude of the Lyapunov exponents. Because the radii of the spheres show only the magnitude of the biggest/smallest Lyapunov exponents but not whether the values are positive, negative or complex, they have to be color-coded by using red for positive (plate 2), green for negative (plate 3) and yellow for complex values (plate 4).

Applying this method to the Rössler system shows, that it is very stable as long as the x-coordinate is negative. When the x-coordinate becomes positive, the system looses more and more of its stability. This method gives a good impression of the local stability of a system. But there is no information of the direction of convergence/divergence.

Visualizing Lyapunov exponents and eigenvectors. As mentioned in section 2 the eigenvectors which belong to the Lyapunov exponents (i.e., the eigenvalues of the Jacobian matrix) denote the direction of convergence/divergence. So it makes sense to display the eigenvectors for several points of the trajectory. The eigenvectors are visualized as lines or tubes. The eigenvectors are scaled with the magnitude of the corresponding eigenvalues (or their real parts if the eigenvalues are complex) and placed so, that the corresponding points on the trajectory is their center. Complex eigenvalues correspond to complex eigenvectors. In this case only the real parts of the complex eigenvectors are displayed. Thus, two complex eigenvectors belonging to conjugated complex eigenvalues coincide, because the real part of two conjugated complex numbers is the same. Again the eigenvectors are color-coded. The eigenvectors of real positive Lyapunov exponents are colored red, the eigenvectors of negative exponents are colored green. The eigenvectors of complex Lyapunov exponents are colored yellow if the real-part is positive, otherwise they are colored blue (see plates 5 and 6).

This method gives directional information which is not illustrated by the technique using the spheres.

Visualizing neighbouring trajectories. Another method for visualizing stability is to observe the evolution of points along neighbouring trajectories. In this case points are positioned in the plane normal to the trajectory approximating a circular cross section. These points are iterated for a certain time intervall to approximate their flow. These trajectories are then connected to build a sweep. The deformation of the sweep gives information of local convergence or divergence of neighbouring trajectories, i.e., stability or instability. After some time steps, the sweep has to be renormalized so that the points around the trajectory have the same distance as before. Without renormalization, the error would increase so much, that the result could not be interpreted any longer to correspond to local behavior. Such a series of short and more or less distorted sweeps are displayed in plates 7 and 8.

When looking at the x-y-plane in negative z-direction one has to follow the trajectory of the Rössler system counterclockwise. When x is negative the sweep that starts with a circular cross section is contracted in z-direction. It ends nearly as a flat plane (plate7). When x becomes positive the sweep is expanded in z-direction which gives an elliptic cross-section (plate8).

Visualizing trajectories of the local linearized system. As we calculated trajectories of the Rössler system (see Plate 1) it is also possible to calculate trajectories of the local linearized system to see the evolution of small perturbations (see section 2.3). As displayed in plate 9, we calculated several trajectories of small perturbations at several points on a trajectory of the Rössler system. The magnitudes of the perturbations are the same at every point of the Rössler system. Also the number of iterations steps remains constant. So one gets an impression of stability and/or rotation at different points of the Rössler system.

Dynamical systems with dimensions greater than three. To apply the obove methods to systems of dimensions greater than three, projections of phase space into 2-dimensional or 3-dimensional space could be done. Displaying several projections simultaneously gives then an impression of the corresponding higher dimensional phase space. Color coding might be used as well to encode, for example, one additional dimension or variable of the underlying dynamical system.

4. Implementation

Our implementation uses the AVS visualization system [AVS92]. AVS allows to build up graphical networks (Fig. 3 left, Fig. 4 left). A network consists of several modules which can generate, convert or output data. In AVS, modules are split into 4 groups: data input modules, mapper modules, filter modules and data output modules. A module may have one or more input and/or output ports.

AVS allows a high degree of interactivity. Each module has its own control panel (Fig. 3 right, Fig. 4 right). It is possible to adjust several parameters for each module interactively. Furthermore automatic changes of parameters can be done to generate animated images.

AVS enables the integration of user defined modules. For implementation of our visualization techniques, two modules named "roessler lyapunov" and "roessler flow" have been written.

4.1. Visualizing Lyapunov exponents

For visualization of Lyapunov exponents, the module "roessler lyapunov" was created. It has two output ports, one for displaying spheres (plates 2-4) and one for displaying eigenvectors (plates 5, 6). One of the output ports is connected to the module "geometry viewer", which is a main visualizing module of AVS (Fig. 3 left). There is also a module "generate axes", which generates coordinate axes. The axes are then converted from simple lines into solid tubes which is done by the module "tube". This conversion makes it possible to render the coordinate axes as solids.

The control panel for the module "roessler lyapunov" allows to adjust the following parameters (Fig. 3 right):

a, b, and c: .... the parameters of the Rössler system
start. x, start. y, start. z: .... the initial point of a solution of the Rössler system

points: .... number of points to display
init: .... number of points to be calculated before data output starts
density: .... distance between neighbouring points
lyap. scale: .... scaling factor for displaying the Lyapunov exponents.
positive/negative/complex: ... which Lyapunov exponent should be displayed (radio buttons)

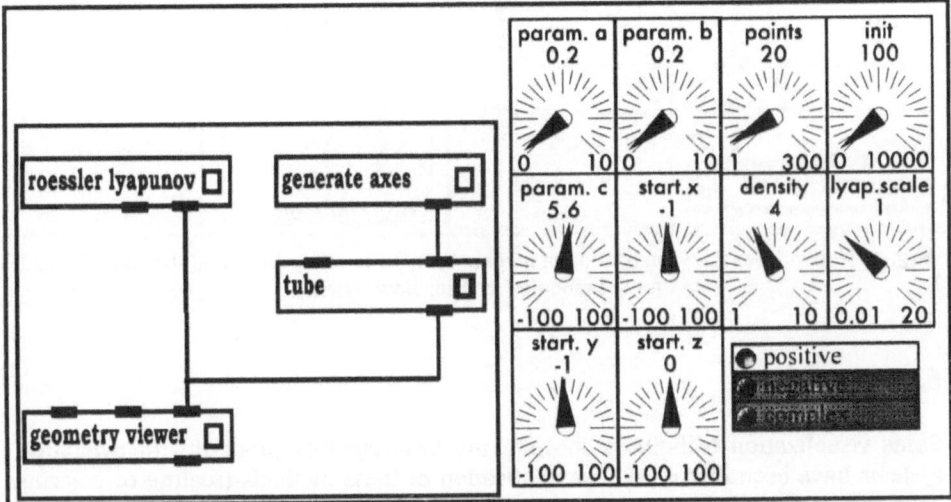

Fig. 3. Network for the visualization of Lyapunov exponents (left) and the control panel for the module "roessler lyapunov" (right).

4.2. Visualizing neighbouring trajectories

The module "roessler flow" is used to visualize the evolution of points on neighbouring trajectories. The output port is connected to the module "field to mesh", which converts a 2D array representation into a polygonal mesh. The module "field to mesh" is connected with the "geometry viewer". The module "generate colormap" is connected with "roessler flow" and "field to mesh" and generates a color table. The module "generate colormap" allows to colorize the mesh. The user may change color assignment according to his needs (Fig. 4 left).

The control panel for the module "roessler flow" allows to adjust the following parameters (Fig. 4 right):

a, *b*, and *c*: .... the parameters of the Rössler system
start. x, start. y, start. z: .... the initial point of the Rössler system
init: .... number of points to be calculated before data output starts
points: .... number of points to display
distance: .... number of points to display before renormalization is done
radius:................................... radius of the circular cross section.

116

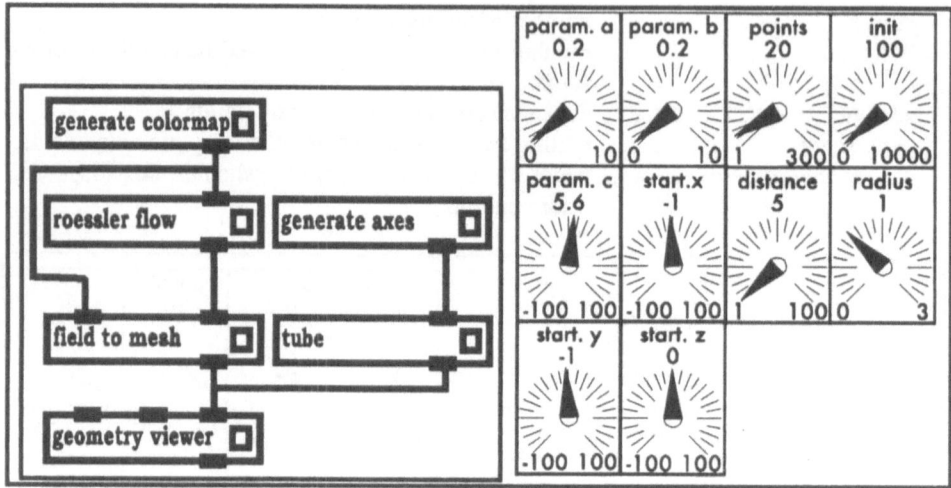

param. a 0.2	param. b 0.2	points 20	init 100
0 10	0 10	1 300	0 10000
param. c 5.6	start.x -1	distance 5	radius 1
-100 100	-100 100	1 100	0 3
start. y -1	start. z 0		
-100 100	-100 100		

Fig. 4. Network for the visualization of neighbouring trajectories (left) and the control panel for the module "roessler flow" (right).

5. Conclusion

Some visualization techniques showing the local stability properties of dynamical systems have been discussed. The application of these methods (scaling or coloring spheres or vectors, investigating the evolution of points on neighbouring trajectories) reveals some interesting details that help to forecast the behavior of dynamical systems.

The Rössler system, which was used as an example of a dynamical system, is a relative simple nonlinear system. With this system it may be possible to recognize the stability properties just by investigating the Jacobian matrix. The presented visualization techniques are, however, applicable to more complex systems as well. They are able to show properties, that are not recognizeable by mathematical analysis only (which might not be feasible at all for complex systems).

6. Future work

The techniques that have been discussed do not show all of the values that have been calculated. In case of local rotation in the presence of complex Lyapunov exponents, there is also a rotation speed and orientation (clockwise or counterclockwise) which might be of interest.

There are also possibilities for volume visualization. Voxels could be color-coded to visualize the stabilities in a large portion of phase space, not just close to a trajectory.

Another way of visualizing stability is to trace sets of randomly positioned points within phase space. It could be observed, whether they are attracted to each other or to a fixed point or if they are repelled from each other.

Instead of a set of points, a plane could be traced along a trajectory in phase space to show several dynamic properties in 2 dimensions.

Different kinds of color-coding can be applied to reveal mixing properties of (chaotic) dynamical systems.

Computer animation is a useful method to show dynamical behavior. With computer animation it may be possible to watch the evolutions of a dynamical system or the evolution of errors in real time.

Acknowledgement. The authors would like to thank Christoph Traxler and Harald Scheirich for proofreading a draft version of this paper.

7. References

[AVS92]   Advanced Visualization Systems Inc., "AVS Developers Guide", Release 4, May 1992.

[Chao89]   "Chaos und Fraktale", Spektrum der Wissenschaft, 1989.

[Fisc94]   Fischel, G., "Visualisierung seltsamer Attraktoren", diploma thesis, Institute of Computer Graphics, Technical University Vienna, 1994.

[Groe94]   Gröller, E., "Application of Visualization Techniques to Complex and Chaotic Dynamical Systems", 5th Eurographics Workshop on Rendering, Rostock, 1994.

[Leeu93]   Leeuv W.C., Wijk, J.J., "A Probe for Local Flow Field Visualization", Proceedings of IEEE Visualization, 1993.

[Lips77]   Lipschutz, S., "Lineare Algebra", McGraw-Hill Book Company, 1977.

[Peit92]   Peitgen, H.-O., Jürgens, H., Saupe, D., "Chaos and Fractals - New Frontiers in Science", Springer Verlag, 1992.

[Schr91]   Schroeder, W.J., Volpe, C.R., Lorensen, W.E., "The Stream Polygon, A Technique for 3D Vector Field Visualization", Proceedings of IEEE Visualization, 1991.

[Tson92]   Tsonis, A.A., "Chaos - From Theory to Application", Plenum Press, 1992.

Editors' Note: see Appendix, p. 157 f. for coloured figures of this paper

Logging in a Computational Steering Environment

Jurriaan D. Mulder*   Jarke J. van Wijk*†

Abstract

Logging of input and output variables is very useful in computational steering. In this paper we describe how we added logging functionality to a computational steering environment developed at CWI. We show how a 2D interface can be augmented with logging by using the third dimension for the display of the logged variables. The user specifies which graphical representations of variables must be logged, and this log is displayed together with the current state of the simulation. Two examples show that logging in computational steering gives more insight in the simulation, that it can be used for monitoring, and that it can be used to undo erroneous actions.

1 Introduction

1.1 Computational Steering

Many new methods, techniques, and packages have been developed for scientific visualization in recent years. Most of these developments however, are limited to postprocessing of data-sets and thus allow no interaction with the simulation itself. Two alternatives to this post-processing approach can be distinguished: *tracking* and *steering* [MKDY90, MH90]. With tracking, visualization is performed during simulation and the only interaction possible is to stop the simulation. With computational steering, simulation parameters can be altered in the ongoing simulation while viewing the results. Such interactive control of a computational model during execution allows the user to quickly discover and correct erroneous input parameter values, but more important, the user gains more insight if he can immediately observe the effect of changes in input parameters to dependent variables.

In [vWvL94] a general and flexible environment for computational steering is described, which has been developed at CWI. Within this environment, the user can easily develop user interfaces and 2D visualizations of his simulation.

---

*Centre for Mathematics and Computer Science CWI, P.O. Box 94079, 1090 GB Amsterdam, the Netherlands. E-mail: `mullie@cwi.nl`

†Netherlands Energy Research Foundation ECN, P.O. Box 1, 1755 ZG Petten, the Netherlands. E-mail: `vanwijk@ecn.nl`

1.2 Logging

Marshall et al. suggest several useful functions for steering applications [MKDY90], one of which is to keep a history of parameter settings. We agree that logging of parameters can be extremely useful in computational steering, especially if such a log can be visualized together with the simulation. In addition, such logging should not be restricted to input variables, but also output variables should be logged. The simulation can take care of the logging, but a more generic and flexible approach is to integrate logging in an environment for computational steering. If a variable can be displayed, it should be possible to log that variable with little effort. As a result, not only the current state of the simulation is visualized, but also an overview of the history can be shown.

A graphical representation of the history of a simulation is useful in several ways. First of all, it provides the user with a better insight in relations between displayed variables. To see how a simulation has developed through time, at one glance or by browsing through its history, can be helpful to understand the results.

Secondly, it dismisses the user of constantly having to watch the simulation progressing. While doing other tasks, the user can check the results occasionally by looking at the visualization of the history of the simulation.

Thirdly, if the history is stored, more functionality can be provided than just its visualization. With a restore option, the user can jump back to the point of time where a special situation occurred and continue the simulation from there, steering it into a different direction. Such a restore option can also be used as an undo function.

Brodlie et al. emphasize the importance of history recording in a problem solving environment that integrates computation and visualization [BB93]. They developed the *history tree* concept, which reflects the history of the search process used by a scientist in reaching an optimal solution to a simulation. This tree is presented to the scientist, who can then perform operations on the tree such as visualize or restore previous simulation data and parameters. Our scope is different from that of Brodlie et al.. Our aim is not to show and handle the history of the process, but the history of the data.

One of the components of the computational steering environment developed at CWI is a graphics tool for the visualization of and interaction with the simulation. In this paper we describe how we extended this tool with logging functionality. In section 2 we give an overview of the computational steering environment. In section 3 we describe the new implementation of logging. Examples are presented in section 4, followed by the conclusions in section 5.

2 The Computational Steering Environment

The computational steering environment (CSE) developed at CWI consists of two major components. A Data Manager takes care of centralized data storage and event notification, and a graphics tool is provided to define a user interface interactively and to show 2D visualizations of the animation. The central concept here is the use of Parametrized Graphics Objects (PGOs): an interface is built up from graphics objects which properties are functions of data in the Data Manager. The architecture of the CSE is shown in figure 1.

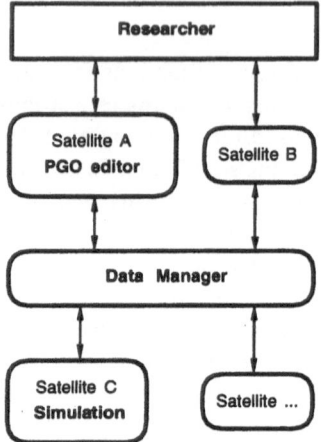

Figure 1: Architecture of the Computational Steering Environment.

2.1 The Data Manager

The central process in the CSE is the Data Manager. Other processes (called *satellites*) such as the simulation and the PGO editor, can connect to and communicate with the Data Manager. The purpose of the Data Manager is twofold. A database of variables is managed, and it takes care of event notification. Satellites can create and read/write variables, and they can subscribe to events, such as notification of mutations of a particular variable. Thus, the Data Manager enables the different satellites to use the same data and to communicate with other satellites.

2.2 The PGO Editor

The PGO editor is a tool for the graphical interaction with a simulation. It handles the visualization of the data, the user input, and direct manipulation. The editor has two modes: specification (*edit*) and application (*run*). In edit-mode, the user can create and edit graphics objects much like in MacDraw-like drawing editors. The standard objects offered are: fill-area, polyline, rectangle, circle, arc, and text. The properties of these objects (such as position and color) can be parametrized to values of variables in the Data Manager. Hence, the user draws a specification of the interface. In run-mode, a two way communication is established between the user and the simulation by binding these properties to variables. Data is retrieved from the Data Manager and mapped onto the properties of the graphics objects. The user can enter text, drag and pick objects, which input is translated into changes of the values of variables.

2.3 Logging Satellite

In [vWvL94] is described how logging can be accomplished with an additional satellite. This satellite keeps track of the changes of one or more variables present in the Data Manager and creates new variables in the Data Manager to store the logged values. With this approach, the logged variables are treated similar as the other variables in the Data Manager. The user must explicitly specify how these logged variables are visualized, in the same manner as the other variables, using the PGO editor.

3 Logging in the PGO editor

The approach we present here is to integrate logging in the PGO editor. The user can select graphics objects and specify that these objects should be logged. The PGO editor then, during run-time, automatically logs the associated variables and displays the logged graphics objects.

Because all graphics in the PGO editor is two-dimensional, we could map the time dimension on the third geometric dimension; the 2D display of the current state is projected on the front plane of a 3D box and the log is displayed inside this box. This way, the log can easily be displayed together with the current state, see figure 2.

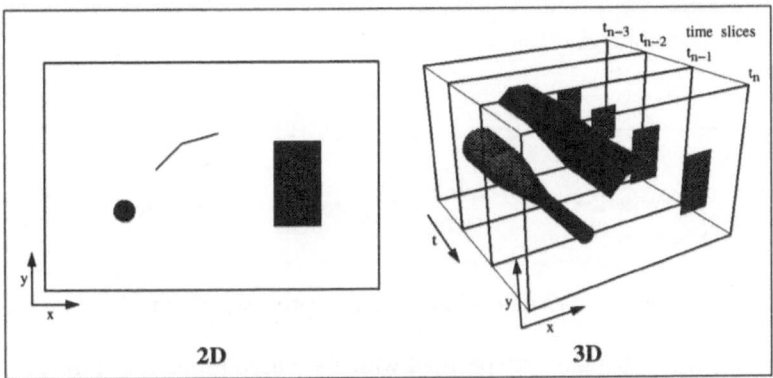

Figure 2: 2D display of current state and 3D display with time slices.

The logged graphical objects can be displayed in two ways: as discrete slices and/or connected in between the slices. With the first display method, the graphical objects are shown in the time slices in the same way they were displayed in the front plane. In the second method, the logged graphical objects are connected in between consecutive slices. Thus, points become lines, lines become surfaces, and areas become volumes. In other words, the 2D graphics objects are swept along the time-axis, just like in translational sweeping in solid modeling [FvDFH90].

When visualizing an ongoing simulation, the user can pause the simulation and thereby freeze the display. The user can then browse through the logged history of the simulation. He can move a plane, parallel to the front plane, along the time-axis. All graphics in front of this plane is clipped. In addition, a restore function is provided. The values of all input variables that are logged are retrieved for the selected time slice and restored into the data manager.

The user can further adjust his view on the box, the number of time steps to be logged, and the logging rate per object.

4 Examples

4.1 Sorting

This first example shows the use of the logging function for the visualization of two sorting algorithms: bubble sort and selection sort. Although our system is not intended for pure algorithm visualization, this example does show how a graphical representation of a log can be useful to understand an ongoing computation.

There are two arrays present in the simulation, each of which is sorted by one of the algorithms. The elements in the arrays can be rearranged into a random or decreasing order. Figure 3 shows the specification of the visualization of the two sorting algorithms. This specification results in a visualization of the two arrays present in the simulation. The elements of the arrays are represented by small rectangles parametrized on the represented value (color) and position in the array (screen position). Just below each rectangle a number is displayed to show the actual value of the element. The number of rectangles that will be drawn in run-mode equals the size of the arrays in the simulation. Two flags indicate whether the arrays are sorted or not.

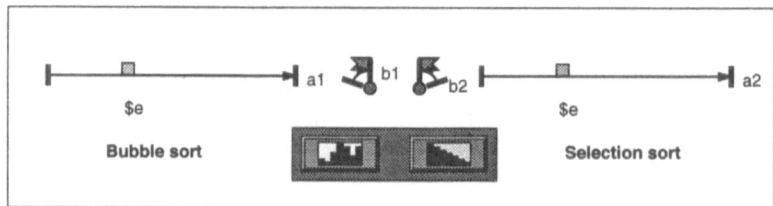

Figure 3: Visualization of sorting process, edit mode.

In addition, two buttons are specified which are used to rearrange the elements in the arrays. If one of these buttons is pressed, the simulation arranges the elements in the arrays to the desired order and starts sorting them. At each exchange in the arrays, the data in the Data Manager is updated, which in turn triggers the PGO editor to update the visualization.

When the sorting process is visualized without any logging, as shown in figure 4, it is not possible to get a good impression of the two sorting algorithms; all there is to see is a number of rapid changes in the arrays until they are sorted. However, if the graphical representation of the arrays is logged, it becomes clear how both sorting algorithms progress. From the displayed log (figure 5) it can easily be derived which elements were exchanged and in what order. Not surprising, the bubble sort algorithm uses far more exchanges than the selection sort.

4.2 Bouncing Balls

Another example is shown in figure 6, where the log of a simulation of bouncing balls is visualized. The balls are subject to a field force, a damping force of the medium, and contact damping in case of collisions. The color of the balls is parametrized to their velocity. The current kinetic energy of the system and the current field force are

123

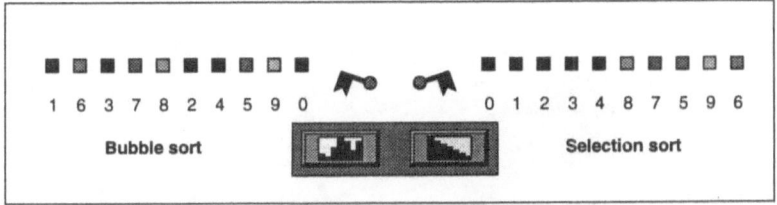

Figure 4: Visualization of sorting process, no logging.

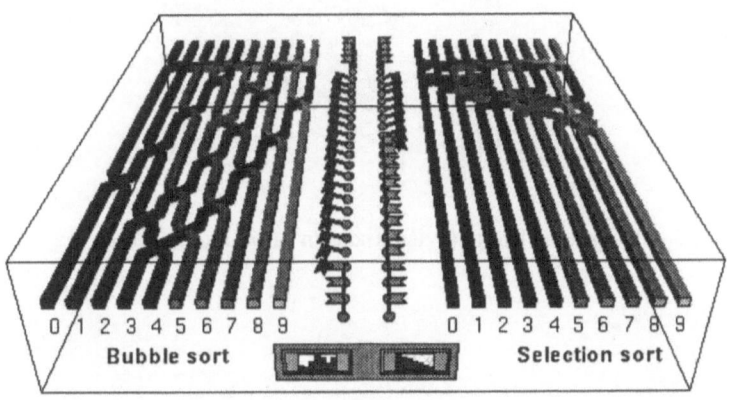

Figure 5: Logged visualization of sorting process.

visualized alongside the field. The user can control the field force and the positions of the balls through direct manipulation.

In figure 6, the balls were grouped together and at rest (a field force of zero). At some point in time, the user invoked a particular field force down the y-axis (a gravity). As a result, the balls accelerated and bounced at the bottom and/or with each other. Due to the damping factors and the field force, the balls slowly come to rest at the bottom.

From the displayed log a good impression is obtained of the acceleration of the balls and the increase of kinetic energy after the activation of the field force. Also, the different trajectories followed by the balls can easily be traced.

In figure 7 the use of the clipping plane to browse through the history of the simulation is illustrated. After pausing the simulation, the plane is positioned just after the point of time where the first collisions occurred. The user could restore the positions of the balls as they occurred at this point of time and continue the simulation, for instance with a different field force.

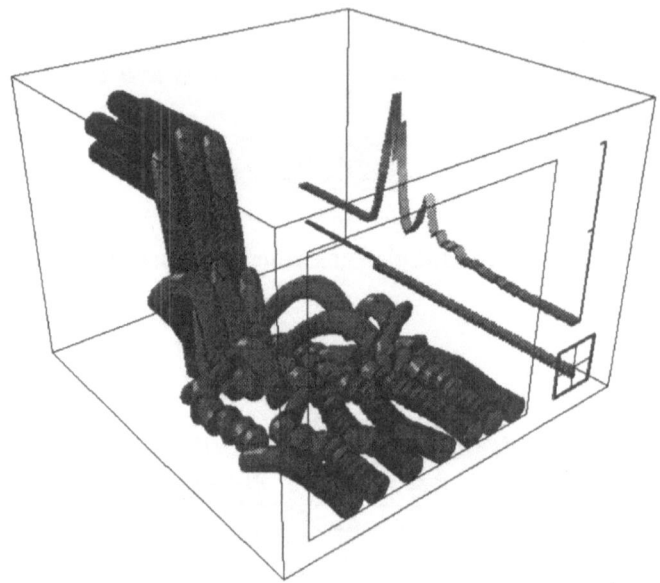

Figure 6: Logged visualization of bouncing balls.

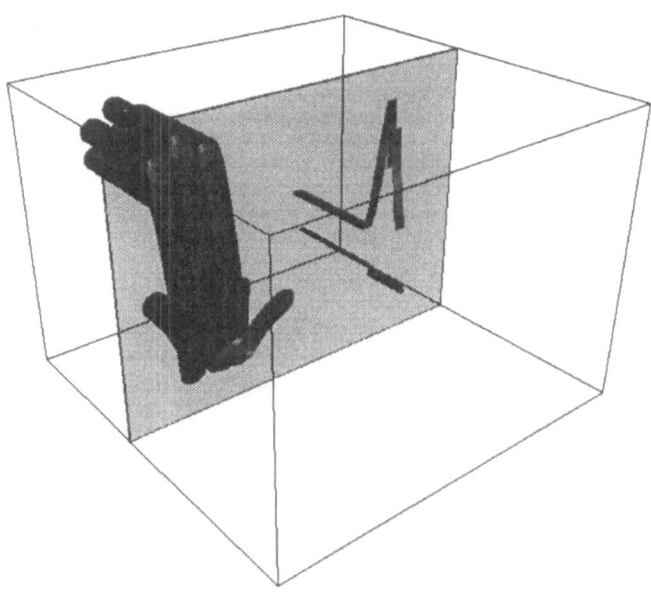

Figure 7: Example of the use of the clipping plane.

5 Conclusion

Logging is an important feature in computational steering. We have extended the graphics tool of the computational steering environment developed at CWI with logging functionality. Since this tool is limited to 2D visualizations, we could use the third dimension as the time-axis and display the logged variables similar to their representation in the current state. More in general, we showed how a 2D user interface can be augmented with logging by using the third dimension for the display of previous states of the interface. We have presented two examples which show the advantages of logging in computational steering: it provides the user with a better insight in relations between variables (an *enhancement* function), it dismisses the user of constantly having to watch the simulation progressing (a *monitoring* function), and it enables the user to jump back to a certain point of time (an *undo* function). However, because this method is only suited for 2D visualizations, the issue on how such a log should be displayed in case of 3D visualizations is still open.

Acknowledgements

The authors would like to thank Robert van Liere (Center for Mathematics and Computer Science, CWI) for his support during the work described in this paper.

References

[BB93]   K. Brodlie and L. Brankin. GRASPARC - a problem solving environment integrating computation and visualization. In G.M. Nielson and D. Bergeron, editors, *Proceedings of the Visualization '93 Conference*, pages 102–109, 1993.

[FvDFH90] J. Foley, A. van Dam, S. Feiner, and J. Hughes. *Computer Graphics: Principles and Practice*. Addison-Wesley, second edition, 1990.

[MH90]   S. Moini and J. Hallquist. Use of concurrent computation and visualization for computation steering in engineering research. In E.J. Farrell, editor, *Extracting Meaning from Complex Data: Processing, Display, Interaction. Proceedings SPIE, 1259*, pages 272–278. The International Society fo Optical Engineering, 1990.

[MKDY90] R. Marshall, J. Kempf, S. Dyer, and C.-C. Yen. Visualization methods and simulation steering for a 3d turbulence model of Lake Erie. *Computer Graphics*, 24(2):89–97, 1990.

[vWvL94] J.J. van Wijk and R. van Liere. An environment for computational steering. Technical Report CS-R9448, Centre for Mathematics and Computer Science (CWI), 1994. Presented at the Dagstuhl Seminar on Scientific Visualization, 23-27 May 1994, Germany, proceedings to be published.

Editors' Note: see Appendix, p. 159 for coloured figures of this paper

Visualization of Internal Combustion Simulations in a Modular Environment

Andrea O. Leone, Riccardo Scateni

CRS4 - Center for Adv. Studies, R&D in Sardinia
Scientific Visualization Group
Via N. Sauro 10, 09123 Cagliari, Italy

Abstract. We describe here a solution to the problem of visualizing the results of simulations of a combustion chamber in a power plant. We used for this a modular visualization environment: Iris Explorer.
We sketch first the fluid-dynamics problem to solve, and then we focus our attention on how to face the visualization problems, especially how to visualize several different scalar fields at the same time.
Then we describe our proposed environment for the solution with the description of several new modules for Iris Explorer we implemented describing their advantages and disadvantages.
Finally, we talk about the possible future evolutions of the project.

1   Introduction

Realistic and reliable prediction of the level of pollutants emitted by power plants burning oil derivates are assuming more and more importance for the choice and the design of future power plants.

While it is still early to try to perform direct numerical simulations (DNS) of turbulent combustors under realistic conditions, a useful approximation can be obtained by reducing the computational grid dimensions using time averaged Navier-Stokes equations.

The type of computations performed to solve these systems are similar to the ones carried out in aeronautics to predict the external flows around reentering space vehicles.

It is quite common to reach dimensions $O(10^2)$ for each component, thus, the number of cells is $O(10^6)$ and for each cell there are, at least, the three components u, v, w of velocity $\vec{V}$, temperature T and a vector of concentrations $\vec{Y} = (Y_1, Y_2, \ldots, Y_{ns})$.

Evaluating the reliability of these computations requires, quite obviously, a large amount of human resources. For this reason a visualization system tightly coupled with the simulation system is extremely useful.

2   The combustion chamber

The process within a combustion chamber is an extremely complex physical system, characterized by a tight connection between fluid-dynamics and chemistry, with an highly non-linear behavior and sensitivity to the boundary conditions, such as the geometry of the system.

This means that to get a reasonable control over the consequences of how modifications to the project (e.g., the injectors' position) can influence the performance, it is not enough to monitor global quantities like the oil percentage actually burnt, but it is also necessary to have a detailed knowledge of the structure of the dynamic flow and of the spatial distribution of the chemistry of the system.

The main goal of the scientists that design new plants is to reach the maximum possible efficiency in burning the fuel while producing the least possible amount of pollutants. To reach this goal the following parameters are varied:

- The shape of the chamber and its dimensions,

- The number and the position of the burners on the walls of the chamber,

- The number and the position of the oxygen injectors,

- The flow gauges of fuel and oxygen.

Varying these parameters in the simulation program results in the generation of several different datasets, each one representing the physical situation inside the chamber and the concentrations of many chemical species.

3   Visualization role

Quantities produced by the simulation process, that is scalar fields, are tightly coupled each other. Analysis of such results, should lead to find out correlations among them.

It is a common practice to use visualization as a qualitative analysis tool. In this case the scientist goal requires two or more scalar quantities to be represented at the same time, so the traditional ways to represent scalar fields (e.g., isosurfaces) becomes inefficient towards gaining insight into the phenomenon.

Even complex techniques, such volume visualization, have to be extended to fit the purpose of the project and to allow an optimal analysis of the results of the described simulations.

For all these reasons we decided to tackle this problem using some *traditional* and some *innovative* techniques.

4   System description

We chose to use Iris Explorer as an embedding environment of our visualization system for mainly two reasons:

128

1. To speed up the development process of the system compared to writing a stand-alone application;

2. To have an easily extendible system.

Using an MVE[1], all the basic building blocks of a visualization system are ready to be used. In the best case, building an application could be as easy as connecting existing modules in a single map. In the worst case, only one or more special modules would be written; but it would be, in any case, reusable for further applications.

This was our choice and the following is the description of the modules and maps we wrote.

4.1 *Traditional* techniques

Our approach to the problem was, first of all, to implement in Iris Explorer several *traditional* visualization methods to analyze the reactive fluid-dynamics fields (mainly scalar, but also vector) and try to figure out how these techniques were useful in understanding the data and what their limitations were. Following is a description of the *traditional* methods we implemented.

[**Scalar fields**] Traditional techniques applied to scalar fields held to the implementation of the maps "SLICE OF A SCALAR FIELD" coloring the different parts by the values, and "ISOSURFACE OF A SCALAR FIELDS". They were, of course, straightforward, but very useful when analyzing a single field.

Variations of these maps held to "MULTI-SLICES OF A SCALAR FIELD" and "MULTI-ISOSURFACES OF A SCALAR FIELDS", allowing the user to visualize more than one slice or one isosurface at the time. They give a broader of the field but still applied to a single scalar field.

Although the last two were very useful in understanding complex structure and distribution of a field, they failed the goal to understand correlations among two or more quantities.

Another refinement of the previous maps led to the implementation of the map "MULTI-ISOSURFACES SLICED OF A SCALAR FIELD", where the user was allowed to visualize more than one isosurface of a field at the time and to cut them with a cutting plane.

[**Vector fields**] We implemented two maps: "VECTOR FIELD SLICED" and "VECTOR FIELD SAMPLED", where user was allowed to view slice or subsample of the whole vector field. Vectors were represented with arrows or segments with direction determined by the value of the field in the origin of the arrow, with length fixed or proportional to the module of the field and optionally colored by the value. They are very useful to understand the velocity vector field and are quite easy to implement. They definitely miss the possibility the analyze the relations among the local behavior of the vector field and the correspondent behavior of related scalar fields in the same place.

[1]Modular Visualization Environment.

[Particle path] Another method to explore a vector field is given by the maps "PARTI-CLES" and "STREAMLINES". Both compute particle paths taking as parameters a given set of initial points and a vector field inside the chamber.

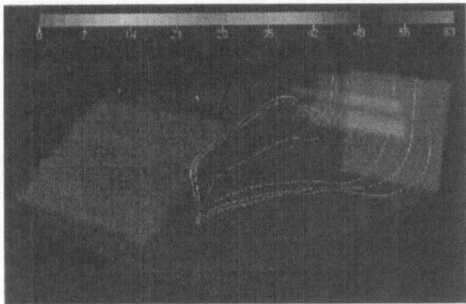

In the first map, small particles are carried in the velocity field, following trajectories determined by the local values of the field. In the second, only the trajectories are computed and displayed.

Both maps are based on a module implementing, as a kernel, a fourth order Runge-Kutta integrator with error estimation performed by time subdivision.

Figure 1: *An example of streamlines of the velocity field.*

The user sets the integration time step and a tolerance value of these two maps by dial widgets. These values are used as control parameters for integration accuracy, with respect to machine precision. Computation of a vector not in a lattice point but inside the chamber, is performed using a tri-linear interpolation method.

Using either of the two, it is possible to evaluate the evolution of the velocity field in the chamber, in some way also simulating the time evolution of the combustion even if the available data are static. Supposing that the situation is steady, it is in fact, reasonable to think that also the vector field, and hence the paths, are steady.

We can see in Figure 1 an example of such a visualization.

4.2 *Non-traditional* techniques

As mentioned in section 3, we implemented non-traditional (or extensions of traditional) visualization techniques to allow analysis of two ore more quantities (chemical species) at the same time, trying to comply with the scientist requests. Here is a summary of these maps implementations, with brief discussions about each map's role.

Scalar-scalar interaction

[Isosurfaces] The map "ISOSURFACE OF A SCALAR FIELD COLORED BY THE VALUES OF ANOTHER SCALAR FIELD" is the direct extension of the map "ISOSURFACE OF A SCALAR FIELD". It gives an idea of the behavior of both the fields but it has the disadvantage that one of them is the *driver* field while the other is the *driven* one. This means that the characteristics of the first field are much more visible than the characteristics of the second.

[Volumetric Visualization] While designing this module we had two goals in mind: represent several scalar fields at a time playing with colors and opacity values, taking advantage of the possibilities offered by the graphics engine of the machines we had available.

To fit in this frame the map "VOLUMETRIC VISUALIZATION" has been implemented by building an appropriate set of triangular meshes with transparency values and colors set as a function of the values in the cell of the scalar fields under consideration. Each cell of the 3D grid is represented by a rectangle (or more precisely two co-planar triangles) with parameterized color and opacity values.

Only one of the three[2] possible sides of the cuboid representing the 3-D cell is visualized: the one with the smaller angle of incidence between the surface normal and the line of sight. This choice is updated while the user manipulates the model in case he/she rotates it.

In this way we thought we could easily render the nebulous appearance of a cloud of chemical elements moving in a combustion chamber. At the same time we could exploit the features offered by the graphics hardware we had available, which was, very good graphics performance in rendering transparent surfaces exploiting of the alpha buffer.

The final result is a scene where a scalar field gives the value of opacity in each cell (it is usually the pressure field) and other fields, from one to three, feed the three color channels, determining in each cell a color that is a valid representation of the blending of the species during the reaction.

In Figure 2 we give an example of such a rendering: opacity represents pressure, the green channel is fed by the concentration of O_2 and the red channel by the concentration of CN.

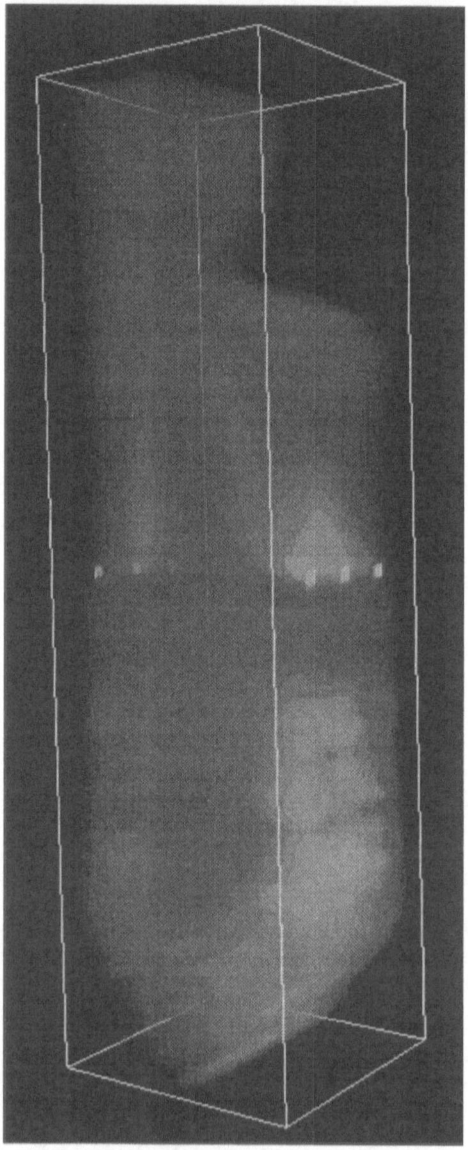

While this representation

Figure 2: *Volume visualization of two chemical species concentration in a pressure field.*

gives almost the same appearance than a ray-casted image of the volume data, the rendering is much faster since the module exploits the graphics hardware of the machine,

---

[2] Actually they are six but parallel each other two by two.

all the object involved are just polygons.

Vector-scalar interaction

[Vector fields] The map "VECTOR FIELDS REPRESENTED AS ARROWS COLORED BY THE VALUES OF A SCALAR FIELD" gives some correlations between a scalar and a vector field and is useful to follow the concentration of certain chemical species in highly turbulent regions, which are easily identified by the behavior of the velocity field. The vector field can be sliced or subsampled as from the maps "SLICED VECTOR FIELD" and "SAMPLED VECTOR FIELD" from which the map derive.

[Particle path] The maps "STREAMLINE COLORED BY THE VALUE OF A SCALAR FIELD" and "PARTICLES COLORED BY THE VALUE OF A SCALAR FIELD" are direct extensions of maps "PARTICLES" and "STREAMLINES" described in section 4.1.

In the first map the positions of the particles are updated according to the vector field and time step integration, and colored by the values that the scalar field assumes in those points. In the second map, trajectories are colored according to scalar field values in the points crossed by the polyline.

Another map implemented is "PARTICLES WHICH RADIUS IS COMPUTED BY THE VALUE OF A SCALAR FIELD". It is a variation of one of the previous, where particles are represented by spheres which radius is computed as function of the scalar field.

All these maps give enough information to understand the relation between the underlying vector field and the selected scalar one.

[Isosurfaces] The map "ISOSURFACE OF A SCALAR FIELD WITH ARROWS FROM A VECTOR FIELD" is an attempt to determine the behavior of a vector field (in our case, the velocity) under certain conditions determined by a scalar field.

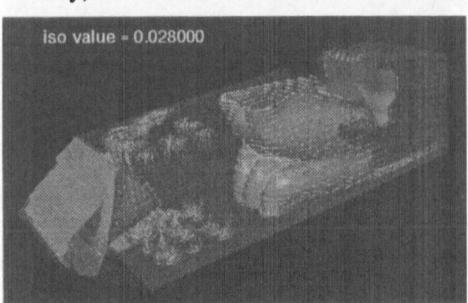

As we can see in Figure 3, the arrows of the vector field are superimposed on the isosurface only in the cells it traverses. For instance, in the case that we are looking at a surface of iso-pressure, we can well identify the rotational effects near the burners since we are only seeing the velocity field

Figure 3: *Vector field superimposed to an isosurface.* were the pressure is uniform on every burner. In other words in this way we are able to identify behaviors otherwise difficult to detect only slicing through the vector field because the surface slicing the volume is not a plane but an arbitrary one. We can thus pretend that this map would perform a sort of *carving* in the vector field determined by a generic surface, in this case the isosurface.

4.3 Deriving new variables

Typically the data to visualize are produced by a simulation program and stored in a file. In this case visualization is a post process. Only saved variable can be showed by the visualization program. Sometimes it is useful to generate derived variables from those saved in the file without overloading the storage process by saving them, performing this task at visualization time. To do this we wrote an Explorer module that computes:

any linear combination of 2 variables Given two variables defined on the same points, and coefficients for them, the module returns a new variable defined on the same points which values are the linear combination of the variable values on that point.

differential operation on a variable Given a variable defined on a set of points, the module returns a new variable defined on the same points which values are a differential operation performed on each point of the given variable. Possible operations are `rot` (curl), `div` (divergence) and `grad` (gradient).

In Figure 4 we can see an example of the value of the divergence of the velocity field computed on the fly while visualizing it. We should also note that this computation is performed (on a multi-processor platform) in parallel with the visualization and thus results in a very small loss of performance compared to the visualization of primitive variables.

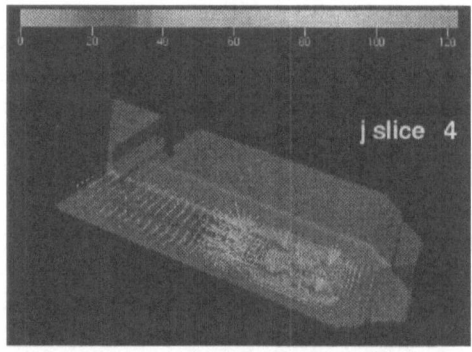

Figure 4: *The vector field of the curl of the velocity computed on the fly by the algebraic manipulator.*

5 Future work

We feel the great need to provide a more friendly support for producing animations, since it is essential to dump a visualization session on tape to discuss the results off-line.

Another important issue to face is the necessity to change the approach to the user interface. As in any 3-dimensional application, it would be of great help to have 3-D tools to manipulate the scene (handles on the geometrical objects, shadow manipulators, 3-d annotation managers etc). At least, 3-D handles would be a further step towards better usability of the system.

An extension to non-regular grids of all the methods presented is also unavoidable if the system is to be as general as possible. Some of the techniques are easily extensible to non-regular grids, whereas the volumetric tool is not so trivial to implement since the concept of minimum angle of incidence is not so easy to apply, for instance, in cylindrical coordinates.

If visualization moves from post-processing to computational tracking or steering of the whole process then the communication must arrange a distributed network of workstations. Even in this case, Explorer would be of great help since is rather easy to interface it with communication tools capable of talking with other processes running on different machines.

Acknowledgements

We would like to thank the people of the Combustion and Microdynamic Simulation Project of CRS4 for their help in understanding the fluid-dynamics aspects of the problems. This work has been financed by ENEL, the Italian power company and the datasets used to generate the paper's figures are supplied by the CRT division of ENEL in Pisa. This work has been also partially carried out with the financial contribution of Sardinia Regional Authorities.

References

[1] Cohen, M. F., Painter, J., Mehta, M., Ma, K.-L. Volume seedlings. Computer Graphics (1992 Symposium on Interactive 3D Graphics), 25(2), 139–145 (1992).

[2] Drebin, R. A., Carpenter, L., Hanrahan, P. Volume rendering. Computer Graphics (SIGGRAPH '88 Proceedings), 22(4), 65–74 (1988).

[3] Grinstein, F. F., Obeysekare, U. R., Patnaik, G. Flow visualization as a basic tool to investigate the dynamics and topology of jets. In: Proc. Visualization '92, pp. 164–170 (1992).

[4] Laur, D., Hanrahan, P. Hierarchical splatting: A progressive refinement algorithm for volume rendering. Computer Graphics (SIGGRAPH '91 Proceedings), 25(4), 285–288 (1991).

[5] Levoy, M. Efficient ray tracing of volume data. ACM Transactions on Graphics, 9(3), 245–261 (1990).

[6] Ma, K.-L., Cohen, M. F., Painter, J. Volume seeds: a volume exploration technique. Journal of Visualization and Computer Animation, 2(4), 135–140 (1991).

[7] Ma, K.-L., Smith, P. J. Virtual smoke: An interactive 3d flow visualization technique. In: Proc. Visualization '92, pp. 46–53 (1992).

[8] Upson, C., Faulhaber Jr., T., Kamins, D., Laidlaw, D., Schlegel, D., Vroom, J., Gurwitz, R., van Dam, A. The application visualization system: A computational environment for scientific visualization. IEEE Computer Graphics and Applications, 9(4), 30–42 (1989).

[9] Upson, C., Keeler, M. V-BUFFER: Visible volume rendering. Computer Graphics (SIGGRAPH '88 Proceedings), 22(4), 59–64 (1988).

134

[10] van Gelder, A., Wilhelms, J. Interactive animated visualization of flow fields. In: 1992 Workshop on Volume Visualization, pp. 47–54 (1992).

[11] van Wijk, J. J. Rendering surface-particles. In: Proc. Visualization '92, pp. 54–61 (1992).

[12] Westover, L. Footprint evaluation for volume rendering. Computer Graphics (SIG-GRAPH '90 Proceedings), 24(4), 367–376 (1990).

Editors' Note: see Appendix, p. 160 for coloured figures of this paper

Visual Simulation of Experimental Oil-Flow Visualization by Spot Noise Images from Numerical Flow Simulation

Willem C. de Leeuw*, Hans-Georg Pagendarm**, Frits H. Post*, Birgit Walter**

*Delft University of Technology, Faculty of Technical Mathematics and Informatics,
Julianalaan 132, 2628 BL Delft, The Netherlands.
**Deutsche Forschungsanstalt für Luft- und Raumfahrt, DLR,
Bunsenstr. 10, D37073 Göttingen, Germany.

Abstract. Comparative visualization of data from different sources provides useful presentations to highlight similarities or differences. Such methods are valuable for comparing results from numerical flow simulation with images taken during windtunnel experiments. The experimental flow visualization technique represents the surface flow field with oil streaks. We visualized the numerical surface flow field using the spot noise technique. The flow data are pre-processed and the parameters of the spot noise texture are tuned to enhance similarity of the resulting images. The result is a 'visual simulation', based mainly on the choice of the quantities to be visualized and the mapping of these quantities to the spot noise parameters. Analysis of the relation between the pre-processing steps and the visualization parameters allows conclusions about the important mechanisms in the experimental flow visualization technique. Besides the comparison of numerical data and the windtunnel experiment, the comparative visualization also provides insight into the visualization techniques involved.

1 Introduction

The issue of comparative visualization has recently received some attention (see Pagendarm and Post [1] for further references). Comparative visualization may be used to compare data from different sources, such as two simulation codes, which address similar physical phenomena. It may be also used to compare visualization methods. Comparison may be performed in various ways, such data level comparison or image level comparison. Results of such comparison may be presented side-by-side as well as integrated in a single image by various techniques. The approach of comparative visualization does not necessarily require images with a high degree of similarity. However, often it is straightforward to tune visualization techniques in such a way that the resulting images look fairly similar. These techniques help the viewer to concentrate on differences in the data rather than needing to integrate the differences resulting from the visualization methods. Pagendarm and Post [1] gave an example of a Schlieren image compared with a numerical flow simulation.

Pagendarm and Walter [2] demonstrated a number of visualization methods on a flow field around a fin/wedge configuration. They also used a photograph of a windtunnel experiment of a similar flow field to compare the data from the numerical solution with the experiment. The visualization of the numerical simulation was not a simulation of the experimental visualization technique. However, the visual similarity of the experimental visualization technique with a texture synthesis technique called *spot noise* became evident. The spot noise technique was developed by Van Wijk [3]. Data from the same case was prepared at DLR in Göttingen to be visualized using an implementation of the enhanced spot noise technique under development at Delft University of Technology. The resulting comparative visualization shows a high degree of similarity. This makes the images easy to interpret by aerodynamicists who are used to such representations from their work at windtunnel facilities.

By varying the data processing steps and the visualization parameters the similarity may be improved. These variations can show the importance of the various factors. Some questions concerning the experimental visualization technique may as well be answered by specific tests with visualization of numerical simulations. This refers in particular to the interesting question whether the oil-flow technique provides local information about the surface flow field at each point on the surface, or whether there exists an integrating effect in streamwise direction.

The paper is organized as follows. In section 2, some background on the flow experiment and simulation will be given. The experimental oil flow visualization technique is described in section 3. In section 4, the data enhancements for the simulated visualization are described. A brief introduction to the spot noise technique, and an explanation of the mapping of the data to the spot noise parameters are given in section 5. In section 6, we present some remarks on the visual comparison, and in section 7 we draw general conclusions.

2 The flow case

The flow field studied here is a simplified configuration of an air-intake of a hypersonic transport vehicle. Figure 1 shows a generic geometry of such an aircraft and its large airduct. The flow condition within the airduct which will contain the engines, will largely determine the performance of such vehicles. A particularly complex region may be found near the corners of the entry section which is expected to have a typical size of 26 m². A complex shock/boundary-layer interaction phenomenon occurs due to the great length of the forebody of the aircraft.

In order to study this complex phenomenon a simplified geometry defined by a blunt fin and a flat plate with a wedge is analysed by numerical flow simulation as well as windtunnel experiments. In the example given here the flow is simulated at a Mach number of 5. The numerical simulation of the compressible flow and the turbulent boundary layers was performed by T. Gerhold [4].

Windtunnel experiments were performed in the Ludwieg tube test facility (RWG) at DLR in Göttingen by P. Krogmann (Gerhold and Krogmann [5]). Hypersonic flow conditions may be established in this test facility for a very short duration of 0.35 seconds. Flow visualization is the main result of such experiments. In this case an oil sublimation technique was used to visualize the surface flow field.

Visualization of the numerical simulation of this flow field has been reported earlier (Pagendarm and Walter [2]). The flow is characterized by several dominant features.

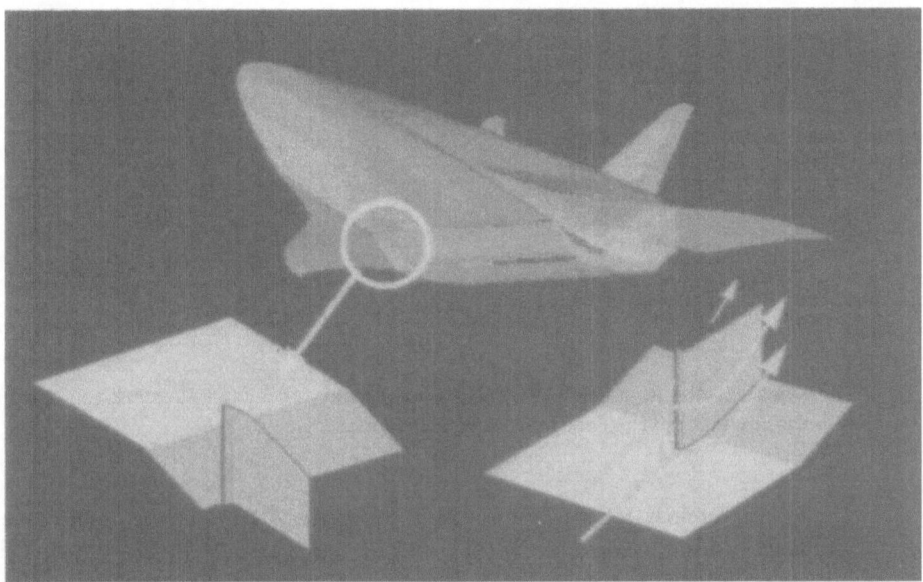

Fig. 1. Schematic illustration of the geometry near a corner of an air-intake of a hypersonic transport aircraft

Horseshoe vortices form in front of the fin and are bent downstream. Shock waves are caused by the fin as well as by the wedge. Both shock waves intersect and interact with the vortices (see figure 2).

3 Surface oil flows

Oil flow patterns have been used for decades for visualization of corner flows which are otherwise hard to access by non-intrusive optical measurement techniques. Colour pigments are mixed with oil and painted onto the surface of a windtunnel model. During the windtunnel test the oil will be blown away or evaporate leaving behind the colour. If a favourable mixture is chosen, the colour forms streak-like patterns on the surface of the model which later will be interpreted by the researcher. Patterns are created because first pigments which stick to the surface tend to accumulate more paint in a small wake behind them. These streaks visualize the direction of the flow close to the solid walls. A more detailed description of this experimental visualization technique was given by Merzkirch [6].

It is the art of the experimentalist to create a pigment/oil mixture in such a way that the oil sticks sufficiently to the wall in order not to drip off. Streaks should be formed by the flow only. On the other hand the oil needs to be blown off or evaporate completely before the steady testing conditions in the windtunnel expire and the flow conditions begin to change. If this is achieved the pattern will remain visible after the flow has stopped. Photographs are used to record the final pattern.

Fig. 2. Two views of the flow field. Vortices are marked by a centered streamribbon and the position of shock waves is visualized by a transparent surface

If the model remains visible from outside the tunnel during testing time, video recording may be used to capture the oil pattern before the flow breaks down. This technique was employed to control the development of the pattern during the test. It is obvious that this visualization technique is not useful for highly unsteady flows.

The technique is extremely useful to determine regions of separation or reattachment which may be identified from converging or diverging streaks. Due to the fact that skin friction is usually much lower near separation the time needed to blow away or evaporate the oil becomes longer. More pigment finds the time to aggregate in the streak and thus the streaks are more clearly visible in these regions.

For the case discussed here, the oil-flow pattern was recorded on a photograph. The streaks clearly indicate the near wall flow direction. Separation is visible on the fin as well as on the plate in front of the wedge.

4 Comparative visualization of experimental and numerical results

Since the numerical simulation is a solution of the Navier-Stokes equations, the flow velocity is zero at the wall. Therefore it is impossible to calculate streamlines on the walls in order to find out more about the surface flow field. A common way of solving this problem is by visualizing skin friction. The wall shear vector τ_w is the derivative normal to the wall of the velocity vector v. In general it is non-zero and points in the direction of the near-wall velocity vectors when projected onto the wall. The wall shear τ_w may be calculated from the gradient of the velocity vector v after introducing a wall coordinate system where n is the coordinate vector of unit length normal to the wall.

$$\tau_w = \nabla \mathbf{v} - (\nabla \mathbf{v}.\mathbf{n})\mathbf{n} \tag{1}$$

The gradient $\nabla \mathbf{v}$ may be calculated using a Gaussian integral formulation

$$\nabla \mathbf{v} = \lim_{V \to 0} \left(\frac{1}{V} \int_A \mathbf{v} dA \right) \tag{2}$$

Fig. 3. The experimental oil-flow visualization (photo: P. Krogmann)

This allows for an easy implementation (Pagendarm and Seitz [7]) that is insensitive to a pathological or degenerated grid cells in the numerical data set. To calculate the gradient, a staggered grid is generated by creating a new vertex at the centre of each grid cell.

After the wall shear τ_w has been calculated on all solid boundaries represented in the data, a standard streamline integration algorithm using a second-order Runge-Kutta scheme is used to integrate the friction lines from the shear vector field. These were a first attempt to compare the experiment with the numerical simulation. For direct comparison, the skin-friction lines were mapped into the photograph of the oil-flow pattern.

In order to create this comparative visualization (figure 4) the camera view parameters in the experimental set-up had to be reconstructed from the photograph. Some information can be drawn from this image. The direction of the wall-shear vector represented by the friction lines is aligned with the direction of the oil streaks for large parts of the image. Some important features in the flow such as the dominant separation line on the side of the fin as well as the separation of the horse-shoe vortex may be found equally in both the experiment and the simulation. The visual appearance of the oil-flow image shows a strong similarity with flow visualization technique as suggested

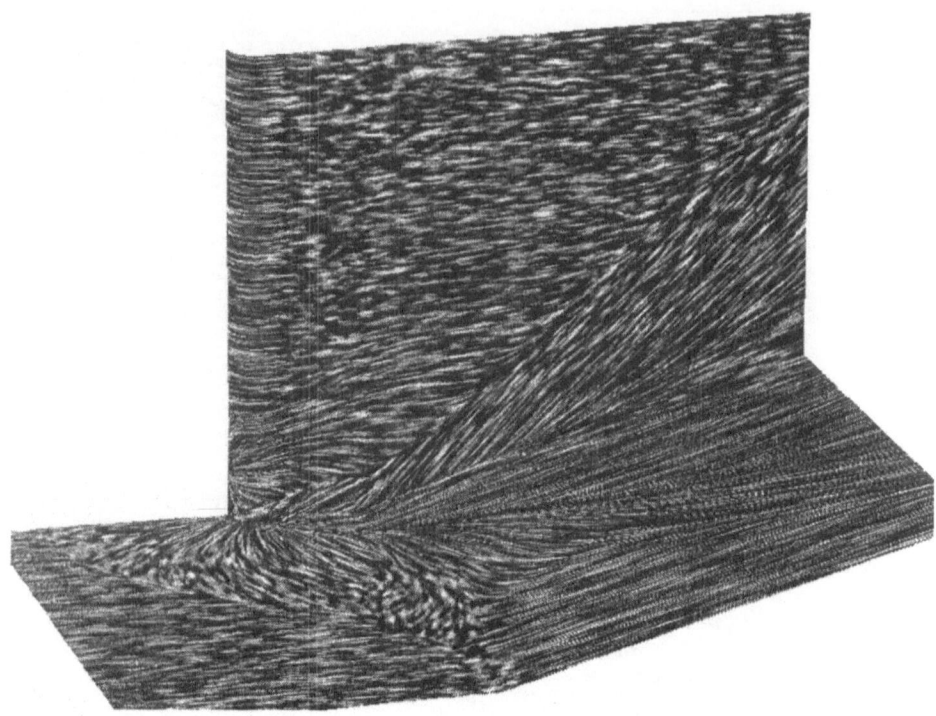

Fig. 5. Spot noise texture generated from wall-friction-vector data

calculated on a curvilinear grid. From earlier experiences with implementing a gradient operator (Pagendarm and Seitz [7]) and a curl operator (Pagendarm and Walter [2]) a Gaussian integral formulation was again used to implement the divergence operator.

$$\nabla.\tau = \lim_{V \to 0}(\frac{1}{V}\int_A \tau.dA) \tag{3}$$

Obviously in discrete space the integration will be performed around a finite volume which is given by a staggered grid defined by the center points of all grid cells. Thus for each node of the numerical grid the (typically) 8 neighbouring cells contribute to the divergence calculation. In order to check whether the convergence $-(\nabla.\tau)$ is a useful quantity to control the spot noise texture, the resulting scalar field was displayed using pseudo-colours (figure 6).

There is a zone of high convergence marked in red at the front of the fin where the cylindrical nose joins without a smooth fairing with the planar side walls of the fin. This was found to be an artefact in the data due to a too low resolution of the CFD simulation. The effect is limited to a small number of grid cells in the proximity of the wall near the kink and is not important for the overall result. To test for spatial correlation, the colour coded convergence data was mapped onto the oil-flow photograph by a simple blending technique.

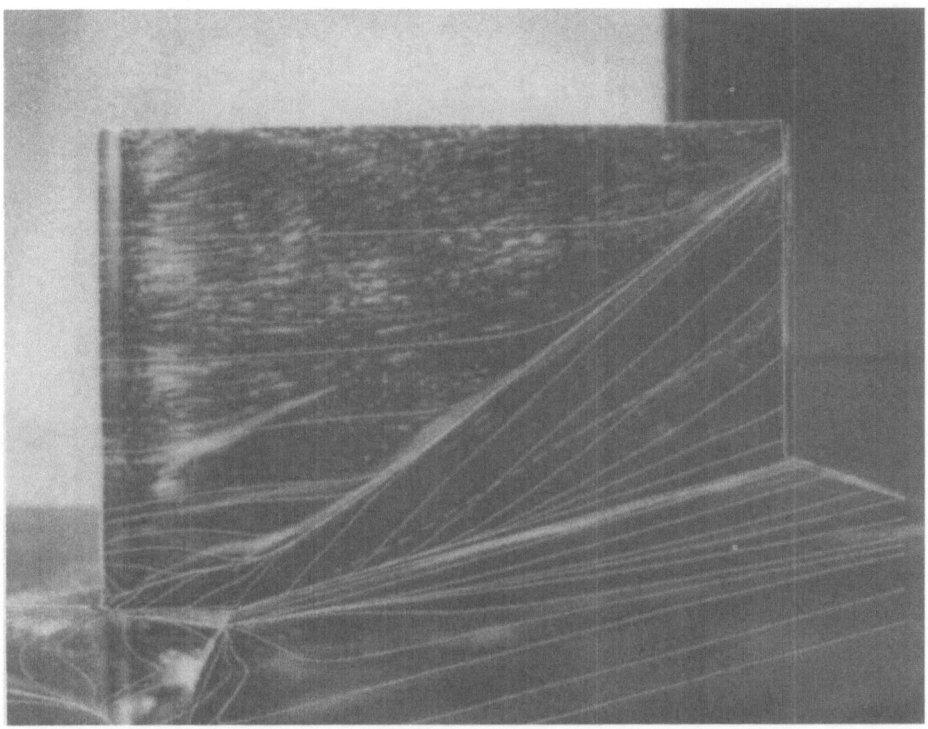

Fig. 4. Superposition of skin-friction lines from the numerical simulation and experimental oil-flow pattern (image reprinted from Pagendarm and Walter 1994)

by Van Wijk [3] using the so-called *spot noise* technique. A flow visualization tool implementing enhancements of this technique has been developed at Delft University of Technology. In a first feasibility study the wall-friction-vector data generated from the results of the numerical flow simulation at DLR were used to generate a spot noise pattern upon the fin/wedge geometry (figure 5).

While the texture of this first study looks promising, it became obvious that additional efforts were required in order to create image which would be as easy to interpret as oil-flow patterns. Looking at the experimental oil-flow visualization (figure 3) reveals that oil streaks are brighter at separation areas. The physical mechanism behind this was explained earlier. Since these areas are characterized by convergent streamlines the divergence (or convergence) of the vector field might be a good candidate to be introduced as a scaling parameter for the spot noise texture.

The convergence of the vector field was selected as a candidate by analysis of the data before actually applying it to the spot noise algorithm. At this early stage the magnitude of the shear vector already turned out to be less useful.

In order to calculate the convergence as quantity DLR's HIGHEND visualization system (Pagendarm [8]) was extended with a divergence operator. Since HIGHEND is a modular system, such extensions are fairly easy to implement. The divergence had to be

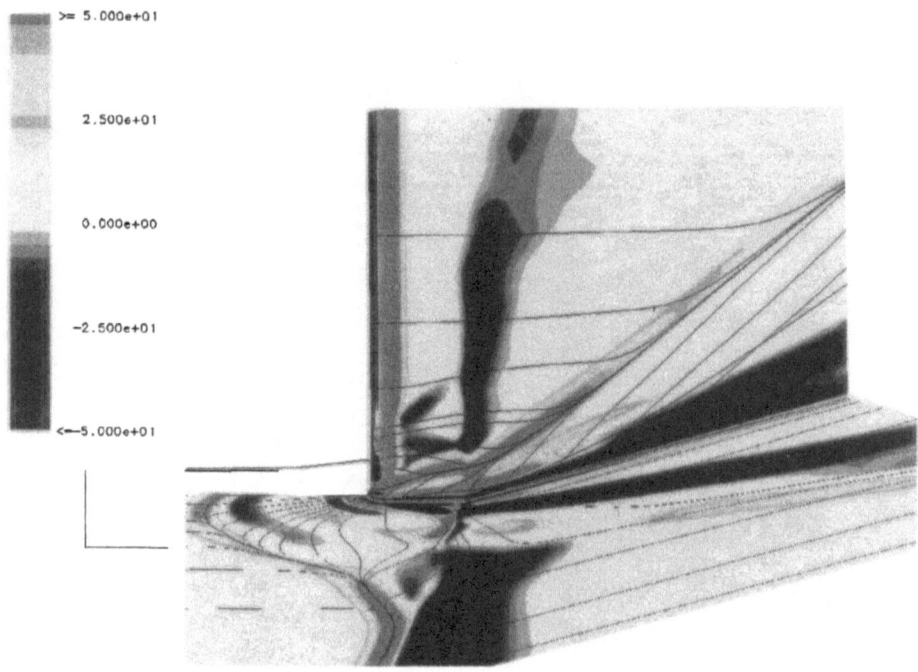

Fig. 6. Colour coded distribution of convergence of the shear-vector field

Figure 7 suggests that the convergence might be an suitable quantity to control the spot parameters. The red, high-convergence locations fit nicely to the thick streaks on the fin and the plate in the oil-flow image which represent the separation position. Now the spot noise technique needs to be examined to find a suitable mapping of parameters in such a way that the texture can represent two different properties of the flow field at the same time. The dominant direction of the spots should represent the direction of the shear vector and the intensity of the spot should vary in some way with the scalar distribution resulting from the convergence/divergence of the vector field.

5 Control of the spot noise texture

Spot noise was introduced by Van Wijk [3] as a texture synthesis technique. The technique is based on a simple local intensity function, called a spot. Large numbers of these spots are placed on random positions on a surface and blended, while the intensity of the spots is scaled by a random value. The result is a texture determined by the basic shape of the spot.

One important use of spot noise is the visualization of 2D vector fields. This is achieved by varying the shape of the spot based on the data at the location of the spot. The spot is stretched proportional with the magnitude of the vector. By simultaneously reducing the width of the spot the area is kept constant. The result is a texture showing

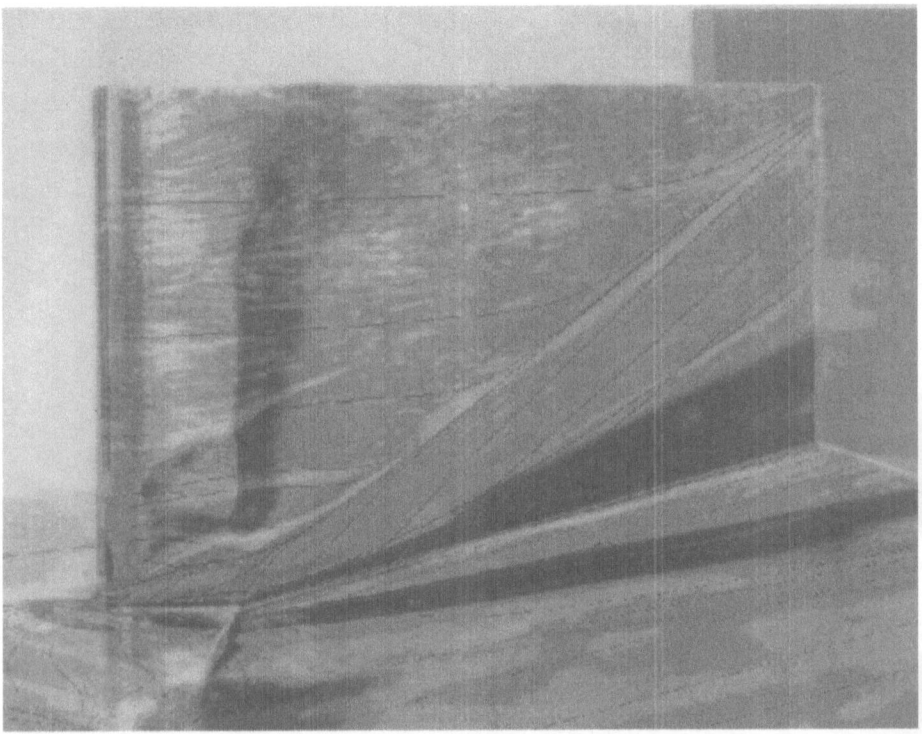

Fig. 7. Colour coded convergence distribution blended with oil-flow pattern

direction as well as magnitude of a vector field. Animation of the textures is possible by treating the spots as particles in a flow. The animation is created by putting the spots at random positions only in the first frame of the animation; for the following frames the positions of the spots are calculated by advection of the initial positions. For visualization of vector fields we use a circular spot, to prevent false impression of direction resulting from using non isotropic spots. The result of using spot noise in its unmodified form is shown in figure 5.

To get a result which can be compared to the photograph of the experiment several modifications were made to the spot noise algorithm. The basic idea was to view the spots used to generate the texture as (white) oil on the (black) surface in the simulation, and thus to emulate the effects of the oil flow on the surface.

First the intensity function was modified. A uniform distribution of positive values was used for scaling of the spot intensity. In standard spot noise a normal distribution with zero means is used for the scaling value for the spot intensity. This improved the perception of the texture as a black surface with white spots on it. A second step was the use of the convergence data (see section 4). Convergence is mapped to the maximum value of the random range that is used to determine spot intensity. In standard spot noise, a constant maximum value I_{max} is used for the random range from which spot intensity I is chosen. We determine the maximum here by multiplying the maximum

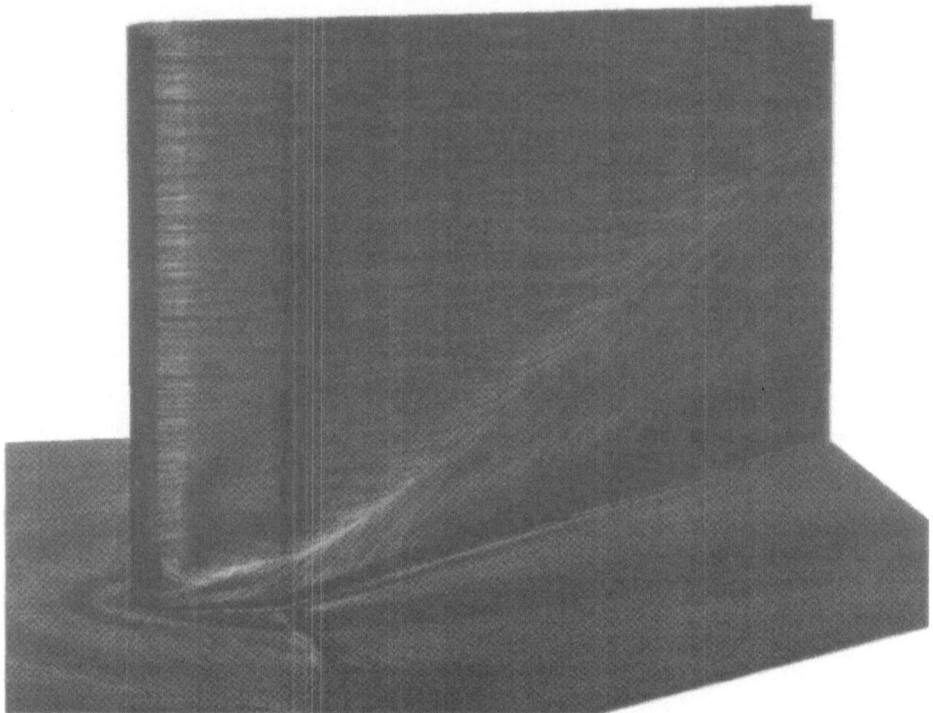

Fig. 8. Convergence data used to scale the intensity distribution function

intensity by the convergence C at the spot position, mapped to the range of 0..1. The intensity I of a spot is thus determined by

$$I = r I_{\max} \frac{C - C_{\min}}{C_{\max} - C_{\min}}, \tag{4}$$

where r is a random value between 0 and 1 and $C_{\min}$ and $C_{\max}$ are the minimum and maximum values of convergence. The effect of using these modifications are shown in figure 8. Although the result resembles the photograph more than the original, some differences are still obvious. These differences are caused by the fact that the technique does not take into account the accumulation of oil in regions of high convergence. To emulate this effect we used 'spot advection'. This means that the positions of the spots were not completely random. Instead, the random positions which would normally be used as the locations to render the spots were advected by the skin friction vector field for a certain time. The resulting positions where used as positions for the spots. The result of this advection is a higher spot density in regions where oil converges. This can be seen in figure 9.

Combining the variable intensity scaling and spot advection techniques leads to our final result shown in figure 10.

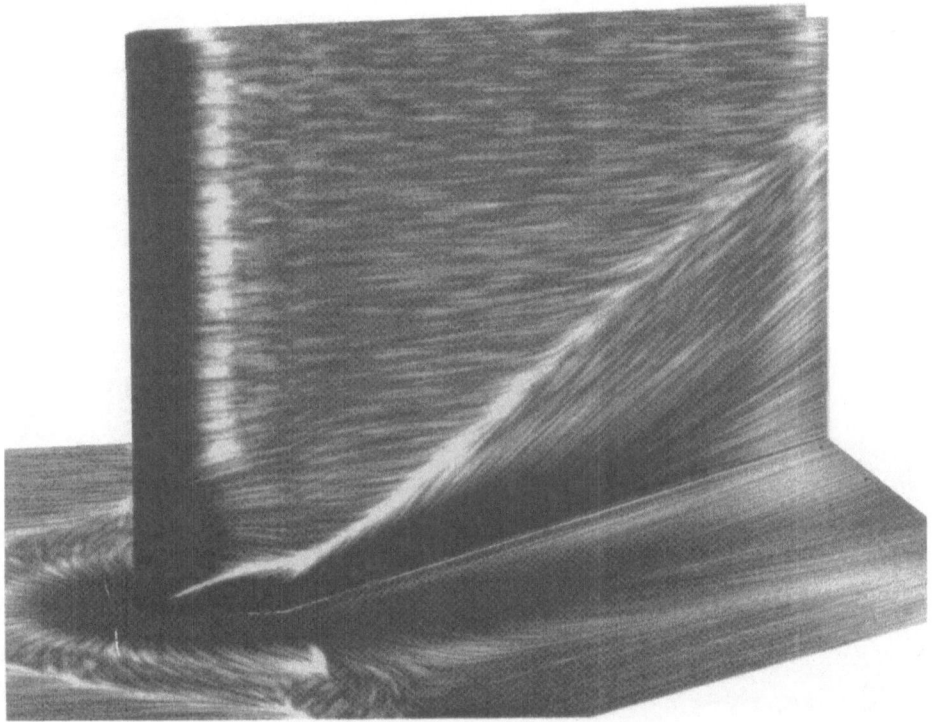

Fig. 9. Spot advection used to simulate oil accumulation in high convergence areas

6 Comparison of spot noise flow visualization with oil-flow patterns

The resulting visualization using the spot noise technique provides realistic images of high quality which are easy to perceive and show a high degree of similarity with the familiar oil-flow visualization of windtunnel tests. The comparison is made at the image level (Pagendarm and Post [1]), and no detailed simulation of the physical oil-flow pattern formation process is needed to obtain pictures that are qualitatively comparable The use of skin friction and convergence data computed from the original velocity field, and of the texture advection technique is based on an informal model of this process, which cannot be considered as a true physical simulation. Both the oil-flow pattern formation and the spot noise generation are essentially constrained random processes, which can be only compared at a global level.

For a closer comparison of spatial correspondence, superposition techniques may be used, but due to the random nature of both visualization techniques at a detailed level, the images do not show differences very clearly. In a side-by-side comparison (figure 11), the most important features of the flow are visible in both images. The geometry of the streak A caused by the main separation is very similar, but the streak B, which is caused by the secondary vortex (Pagendarm and Walter [2]) does not appear in the numerical visualization at all. Detailed analysis reveals that this is caused by a lack

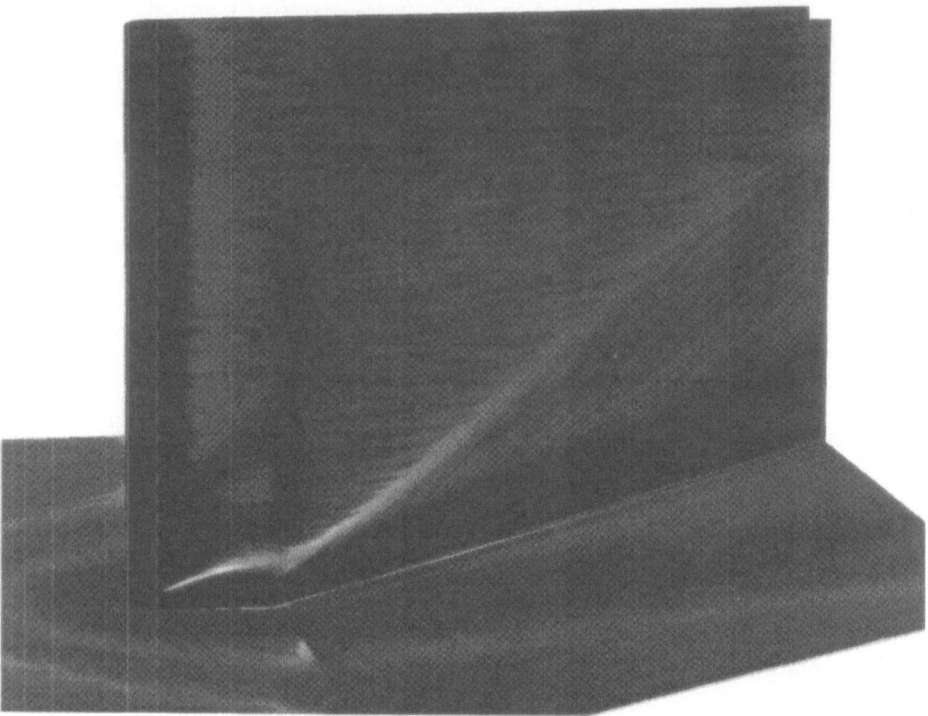

Fig. 10. Final result combining the effects of Figures 8 and 9

of resolution close to the wall in the numerical simulation. The sharp sloping streak E is also absent in the spot noise image. Discussion with the aerodynamicists suggests that the brightness of the oil in this part of the oil flow picture could be a result of an uneven initial distribution of the oil in the experimental set-up. Figure 4 shows that both streaks B and E are not aligned with the skin friction lines. While in the case of streak B this was identified as a deficiency of the numerical simulation, the discrepancies at E are most probably a fault of the experimental conditions. The streak E is located in a region of low friction. The streak is formed during the start-up phase of the windtunnel. Due to the short duration of steady flow condition (< 0.35 seconds), the oil pattern may not fully adjust to the final flow in regions where forces on the oil are low. The faint trace marked by C is a trace of a contact discontinuity in the flow field. This feature is present in the experiment as well as in the numerical simulation. Since the resolution of the numerical mesh decreases rapidly towards the top of the fin, this feature is again represented too weakly in the numerical data. Differences near the top of the fin are due to the fact that the numerical simulation implements a fin of infinite length as a boundary condition. The vertical streak near the nose of the fin at G is a specular highlight; also, a mirror image of the fin in the plate is visible. These optical effects could be easily included in the spot noise image, but this would not be useful for the visualization. The purpose of

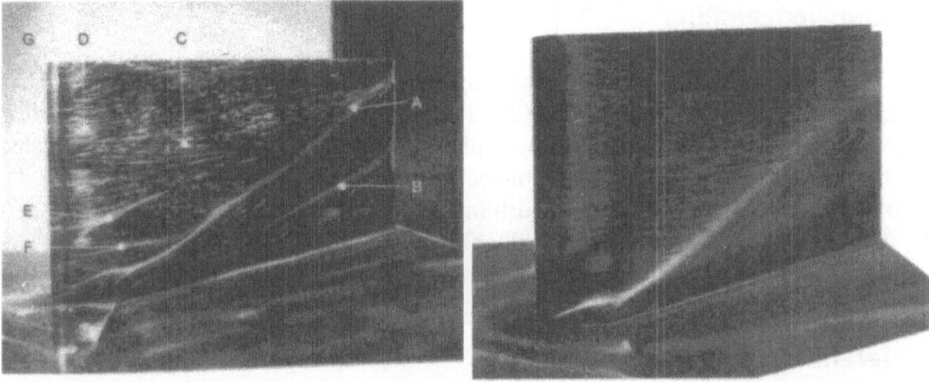

Fig. 11. Side-by-side comparison of experimental and spot noise visualization (figure 3 and figure 10)

realism here is only to achieve good comparability.

A problem of the image level comparison is that viewing of the experimental image from other viewpoints is impossible. A top view photograph of the experiment is also available (Gerhold [4]), and thus a comparative analysis of the complex patterns on the plate is possible. But the number of camera views is necessarily limited, and this again makes the case for a comparison, based on feature extraction, which is less dependent on viewing parameters.

7 Conclusions

We have demonstrated a visual simulation of the oil-flow patterns from a windtunnel experiment by an enhanced spot noise texture, resulting in visually similar images that are well suited for qualitative comparison. A good comparison of the numerical and experimental visualizations is greatly helped by a visual simulation of the experimental process, based only on graphical techniques. The spot noise technique has proved very useful for this type of application.

Further research is needed to see if the visual simulation technique works equally well for other skin friction data sets. Also, a detailed physical modelling of the oil streaking process may produce better results, but such a model would be highly specialized for this type of visualization, and computationally expensive.

The significance of differences found in the comparison can be established by a hypothetical physical explanation. Irregularities may be explained by uneven distribution of the oil on the fin surface, but sometimes also by features of the flow field, as is the case with the secondary vortex. Comparative analysis of the flow simulation and the visual simulation technique may lead to such explanations. The image comparison is a very fruitful source of hypotheses for further investigation, both in fluid dynamics and in scientific visualization.

148

Acknowledgements

This research is a result of the informal cooperation between DLR in Göttingen and TU Delft on comparative visualization in CFD research. Wim de Leeuw's work is supported by a grant from NWO/SION. We would like to thank Jarke van Wijk for his valuable contributions to the research on spot noise, and Remko Vaatstra for his help with the implementation of the spot noise algorithm. The simulation data was kindly supplied by T. Gerhold who also contributed valuable discussion of the physics of the flow. The oil flow photograph is courtesy of P. Krogmann.

References

1. H.-G. Pagendarm and F.H. Post. Comparative visualization – approaches and examples. In M. Göbel, H. Müller, and B. Urban, editors, *Visualization in Scientific Computing*, pages 95–108. Springer, Wien, 1995.
2. H.-G. Pagendarm and B.Walter. Feature detection from vector quantities in a numerically simulated hypersonic flow field in combination with experimental flow visualization. In D. Bergeron and A. Kaufman, editors, *Proceedings Visualization '94*, pages 117–123. IEEE Computer Society Press, 1994.
3. J.J. van Wijk. Spot noise – texture synthesis for data visualization. In T.W. Sederberg, editor, *Computer Graphics (SIGGRAPH '91 Proceedings)*, volume 25, pages 263–272, July 1991.
4. T. Gerhold. Numerische Simulation und Analyse der turbulenten Hyperschallströmung um einen stumpfen Fin mit Rampe. Technical Report DLR-FB 94-19, DLR, 1994.
5. T. Gerhold and P. Krogmann. Investigation of the hypersonic turbulent flow past a blunt fin/wedge configuration. AIAA-93-5026, 5th Intern. Aerospace Plane and Hypersonic Technology Conference, Munich, Germany, Nov.30-Dec.3, 1993.
6. W. Merzkirch. *Flow Visualization (2nd edition)*. Academic Press, 1987.
7. H.-G. Pagendarm and B. Seitz. An algorithm for detection and visualization of discontinuities in scientific data fields applied to flow data with shock waves. In P. Palamidese, editor, *Scientific Visualization – Advanced Software Techniques*, pages 161–177. Ellis Horwood Ltd, 1993.
8. H.-G. Pagendarm. HIGHEND, a visualization system for 3D data with special support for postprocessing of fluid dynamics data. In M. Grave, Y. LeLous, and W.T. Hewitt, editors, *Visualization in Scientific Computing*, pages 87–98. Springer, Heidelberg, 1994.

Editors' Note: see Appendix, p. 161 for coloured figures of this paper

Appendix: Colour Illustrations

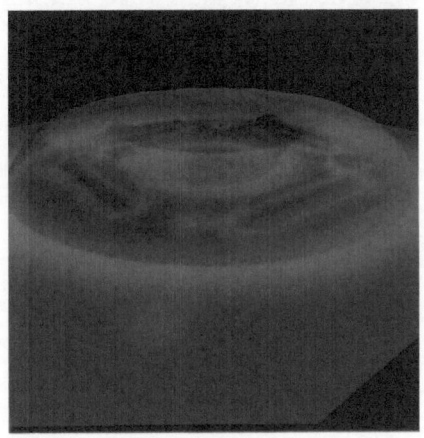

Horizontal planar cut through torus
(Pang and Clifton, Fig. 9)

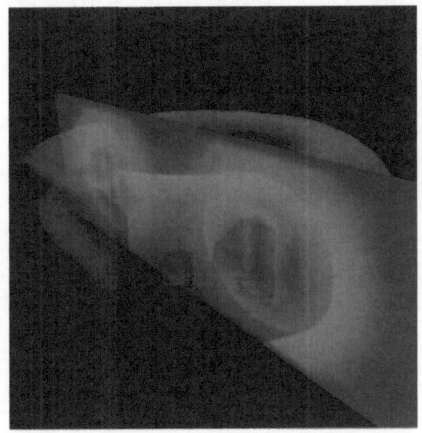

Arbitrary planar cut (Pang and Clifton,
Fig. 10)

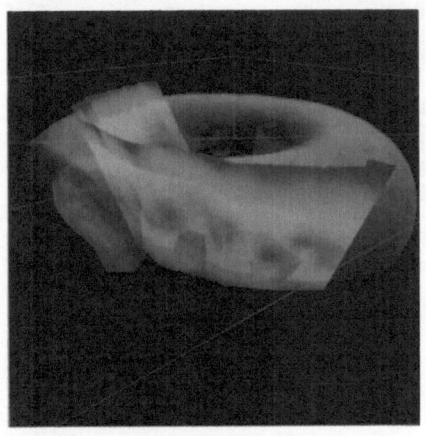

Curved cut through torus (Pang and
Clifton, Fig. 11)

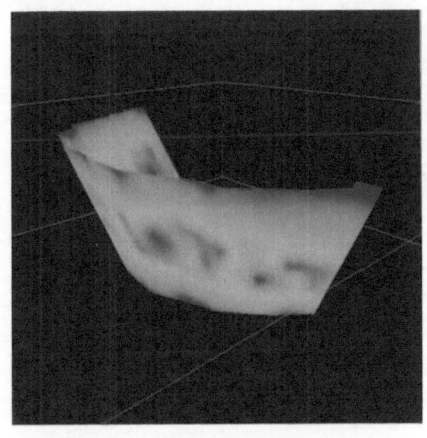

Curved cut alone (Pang and Clifton,
Fig. 12)

150

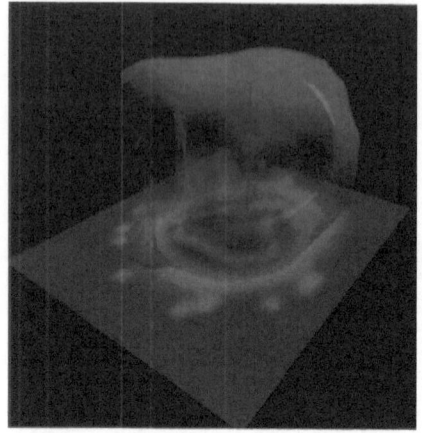

Horizontal planar cut through head
(Pang and Clifton, Fig. 13)

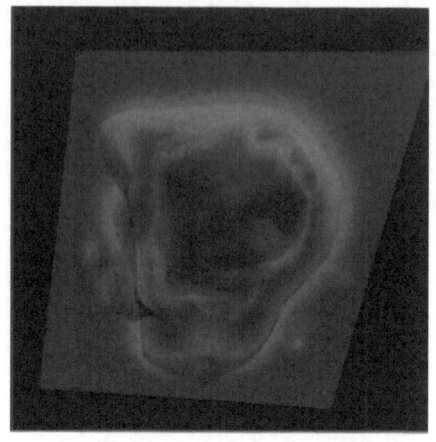

Arbitrary planar cut (Pang and Clifton,
Fig. 14)

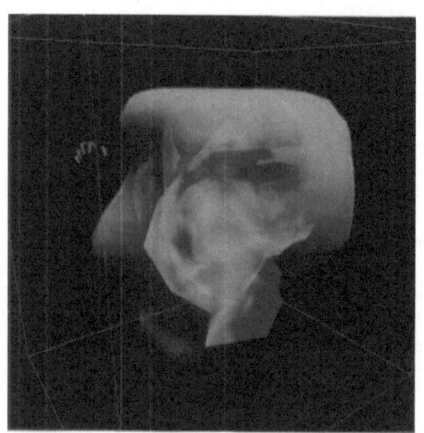

Sweeping curved cut through head
(Pang and Clifton, Fig. 15)

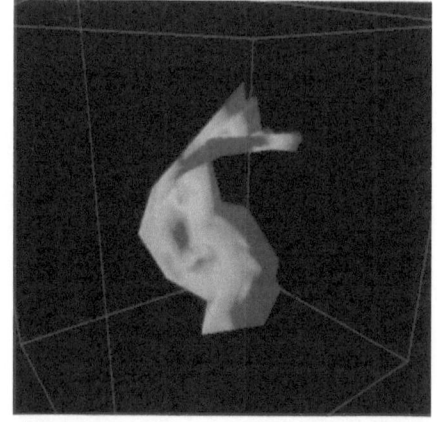

Curved cut alone (Pang and Clifton,
Fig. 16)

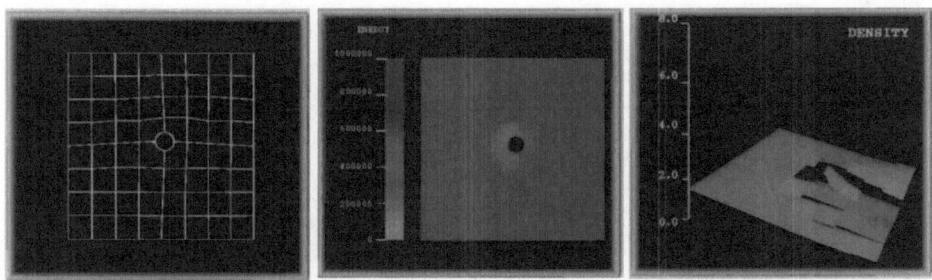

a Grid, color encoded energy and height encoded density values at time 0.3

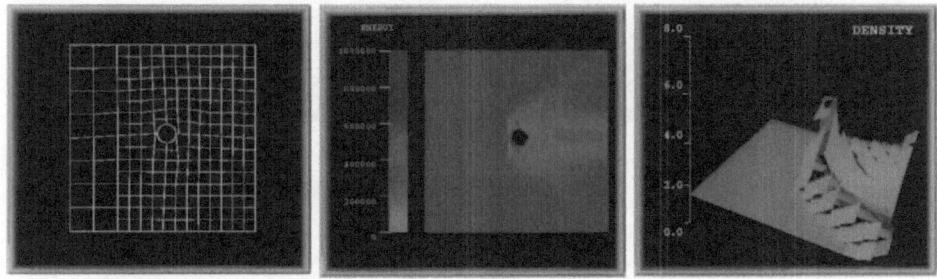

b Refined grid, color encoded energy and height encoded density values at time 0.7

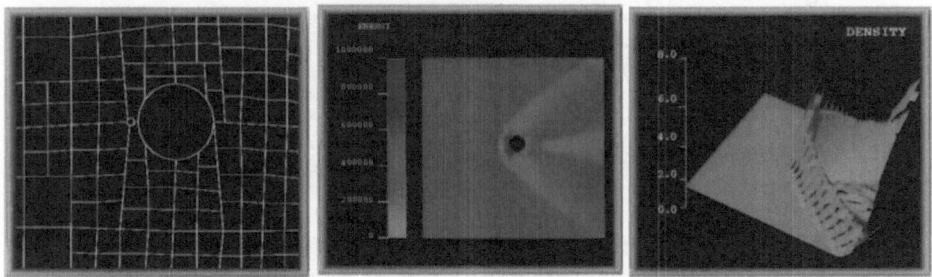

c Detail of further refined grid, color encoded energy and height encoded density values at time 1.0

Example session of an on-line visualization (Schmidt and Rühle, Fig. 7)

152

Fig. 3

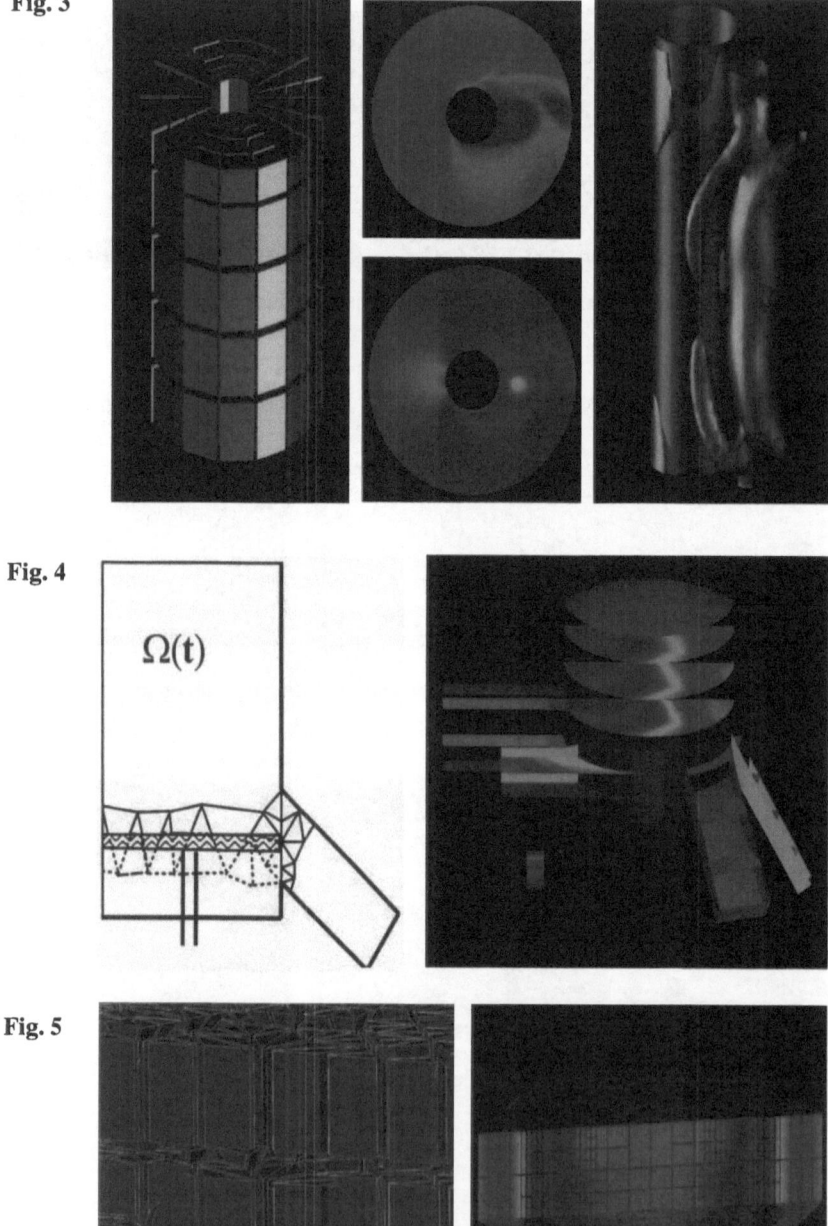

Fig. 4

$\Omega(t)$

Fig. 5

(Figs. 3–5, Rumpf et al.)

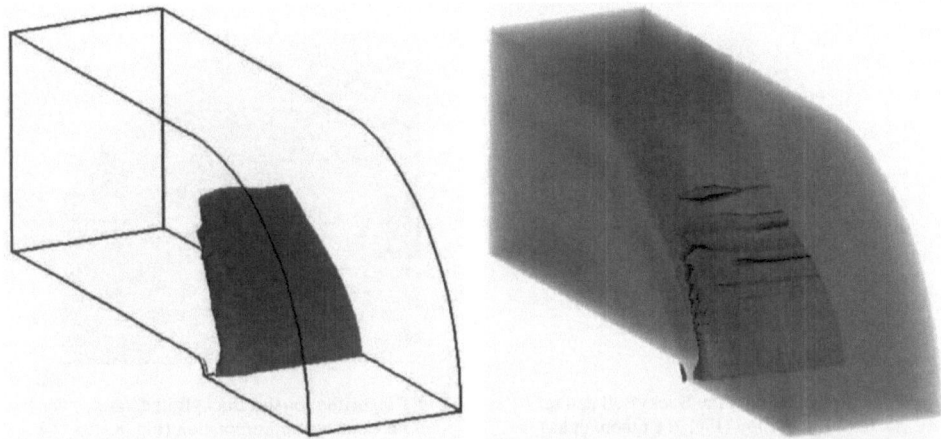

The Bluntfin dataset (Density Isosurface at 1.745). Left: Isosurface via 'Marching Cubes for curvilinear grids'. Right: Combined semi-transparent/opaque raycasting (Frühauf, Colour Plate 1)

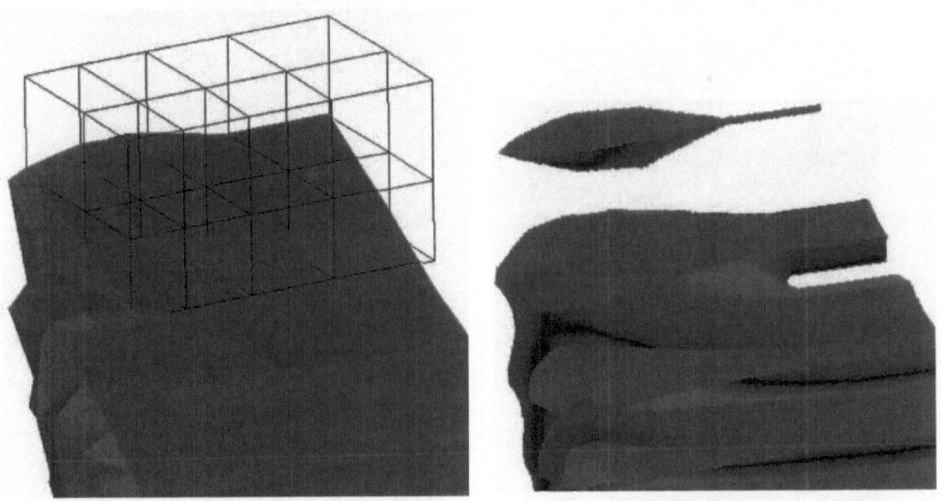

Isosurfaces in the Bluntfin dataset (Density = 1.745). Left: Isosurface via 'Marching Cubes for curvilinear grids' (with part of the computational grid). Right: Isosurface via opaque raycasting (Frühauf, Colour Plate 2)

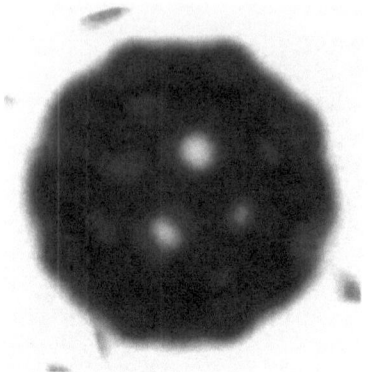

PT algorithm on the BuckyBall dataset
at full precision (Fig. 7, Cignoni et al.)

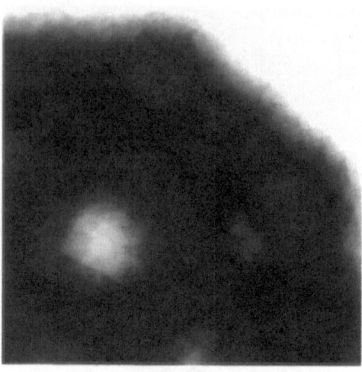

PT algorithm on the BuckyBall dataset,
with Centroid approximation (Fig. 8,
Cignoni et al.)

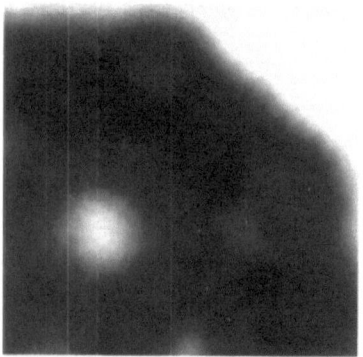

PT algorithm on the BuckyBall
dataset, with UTS2 approximation
(Fig. 9, Cignoni et al.)

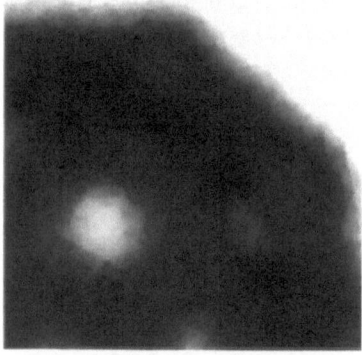

PT algorithm on the BuckyBall dataset,
with Voxel approximation (Fig. 10,
Cignoni et al.)

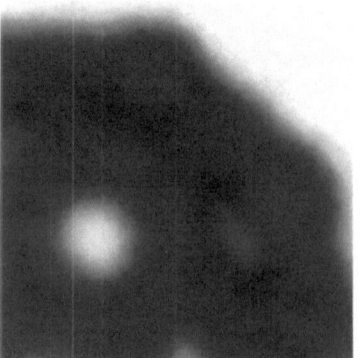

PT algorithm on the BuckyBall dataset
at precision 2% (Fig. 11, Cignoni et al.)

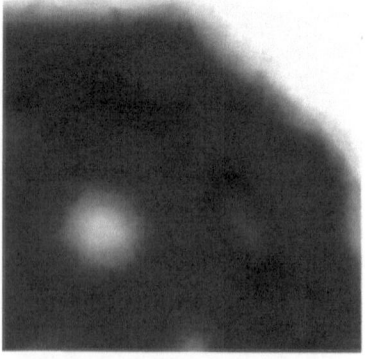

PT algorithm on the BuckyBall dataset
at precision 5% (Fig. 12, Cignoni et al.)

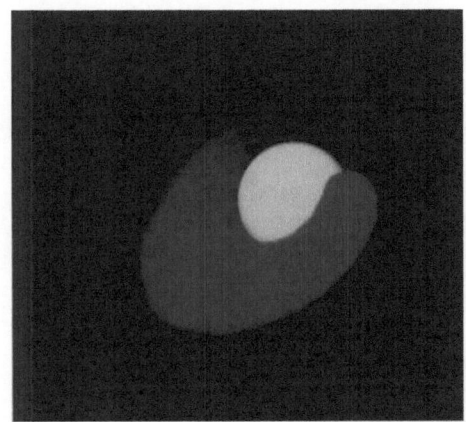

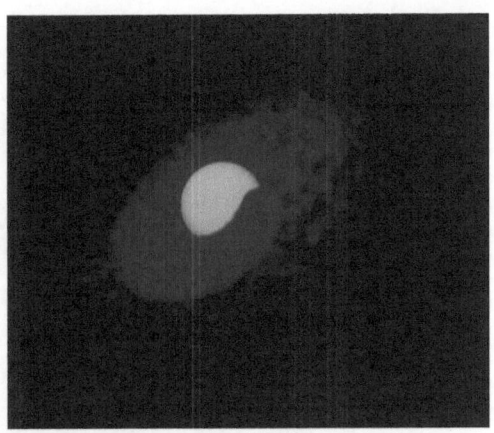

SPH simulation of the formation of a Be star disk. Smoothing length $h \approx 1/30\ R_{star}$. Time = 100. Mass density visualized with smoothing kernel (Rau and Straßer, Fig. 3)

Time = 400 (Rau and Straßer, Fig. 4)

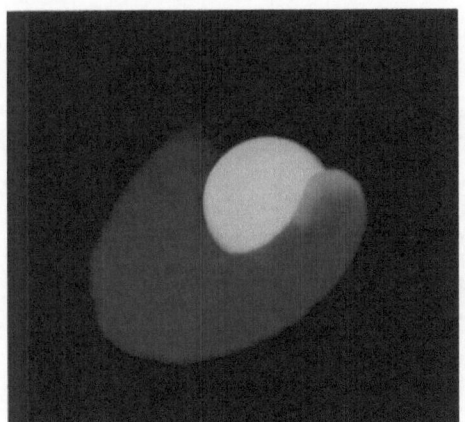

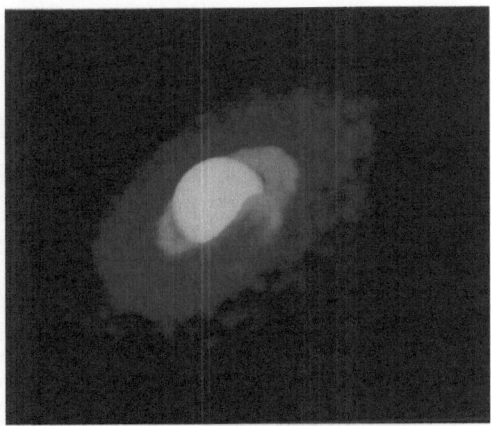

Time = 100. Mass density visualized with smoothing kernel and additional color coding (Rau and Straßer, Fig. 5)

Time = 400 (Rau and Straßer, Fig. 6)

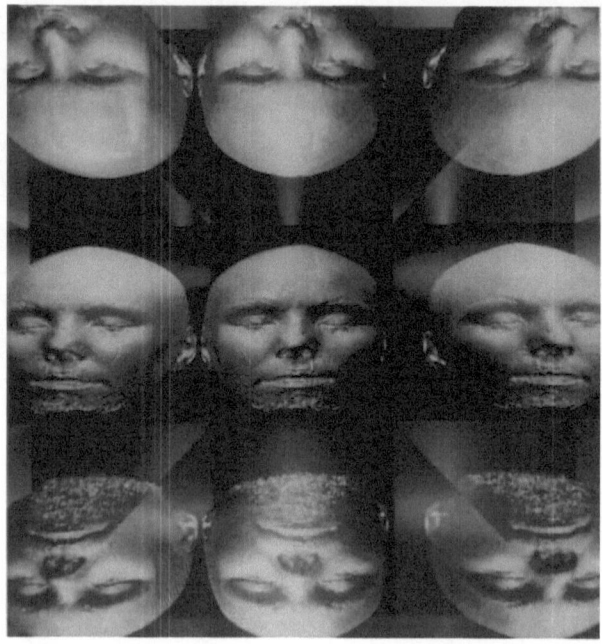

Test scene with the head data (Zhang and Liu, Fig. 5)

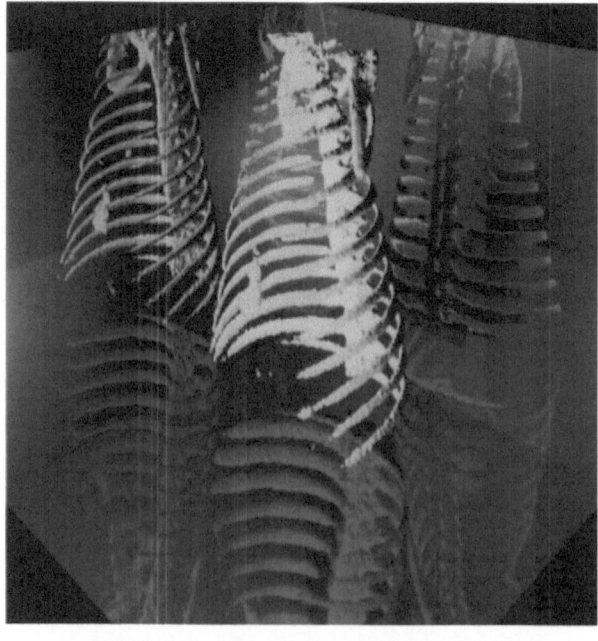

Test scene with the trunk data (Zhang and Liu, Fig. 6)

157

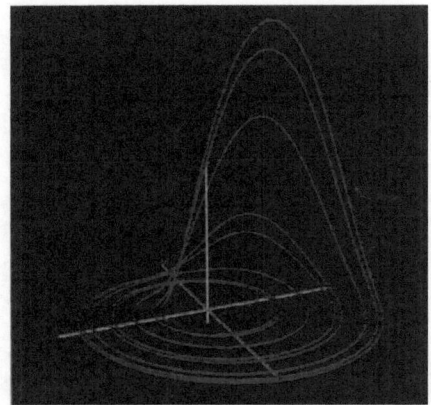
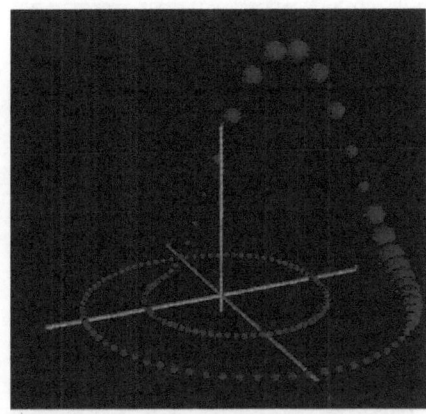

Some trajectories of the Roessler system (Fischel and Gröller, Plate 1)

Spheres showing the largest positive Lyapunov exponents (Fischel and Gröller, Plate 2)

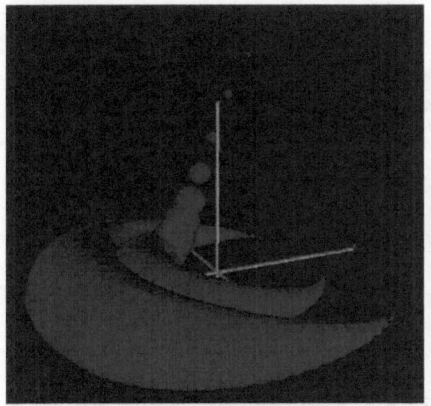
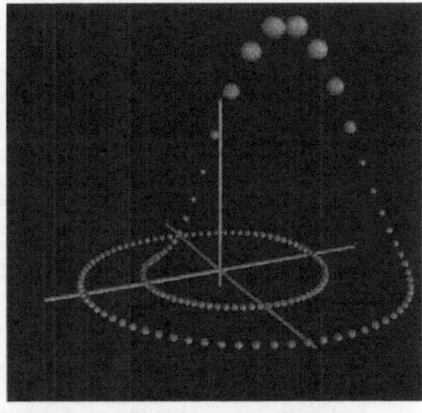

Spheres showing the largest negative Lyapunov exponent (Fischel and Gröller, Plate 3)

Spheres showing the imaginary part of complex Lyapunov exponents (Fischel and Gröller, Plate 4)

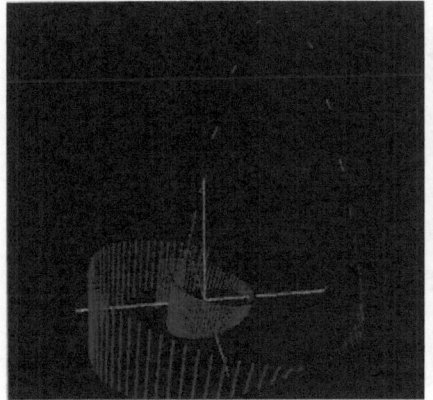

Eigenvectors showing the direction of convergence/divergence (Fischel and Gröller, Plates 5 and 6)

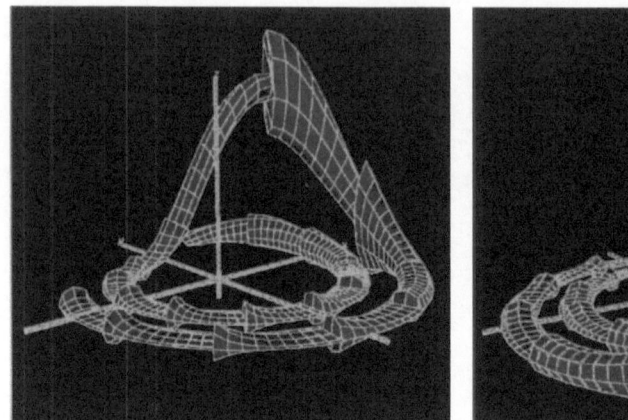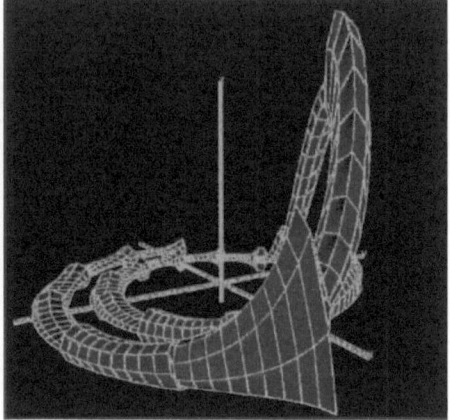

Evolution of neighbouring trajectories (Fischel and Gröller, Plates 7 and 8)

Trajectories of local linearized systems showing the evolution of local perturbations (Fischel and Gröller, Plate 9)

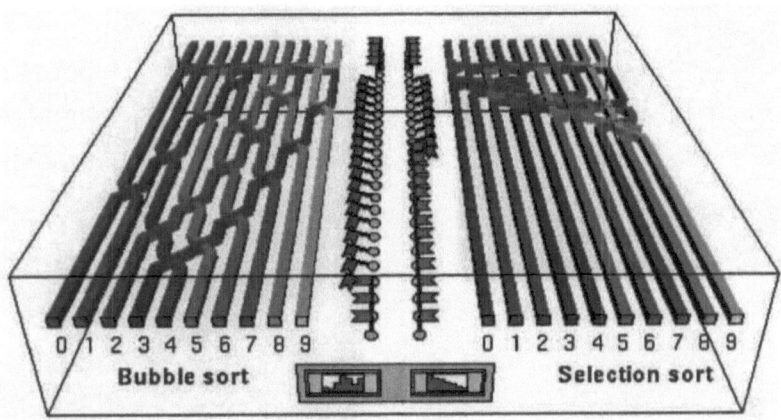

Logged visualization of sorting process (Mulder and van Wijk, Fig. 5)

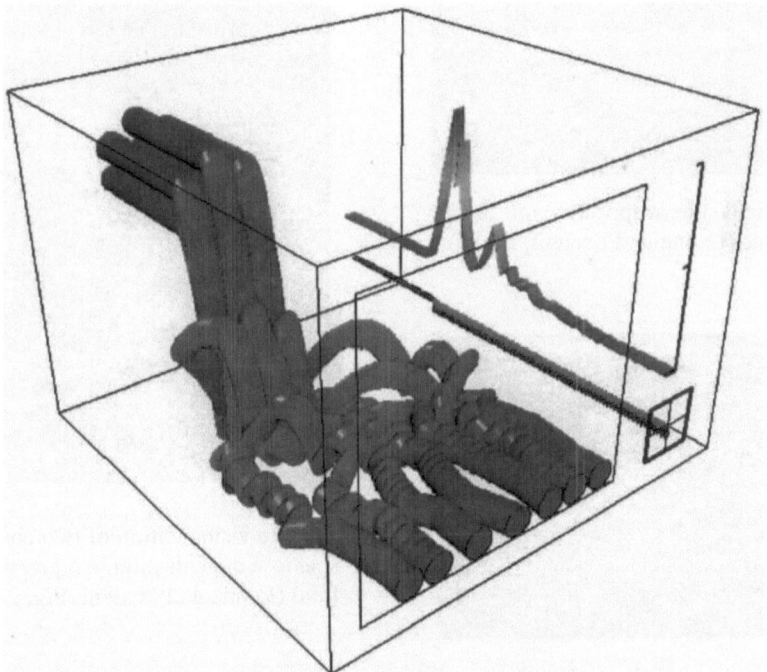

Logged visualization of bouncing balls (Mulder and van Wijk, Fig. 6)

An example of streamlines of the velocity field (Leone and Scateni, Fig. 1)

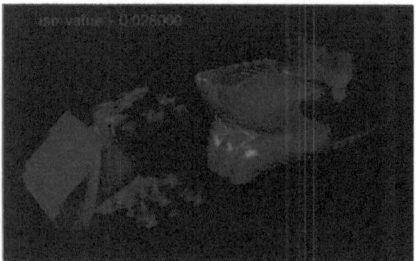

Vector field superimposed to an isosurface (Leone and Scateni, Fig. 3)

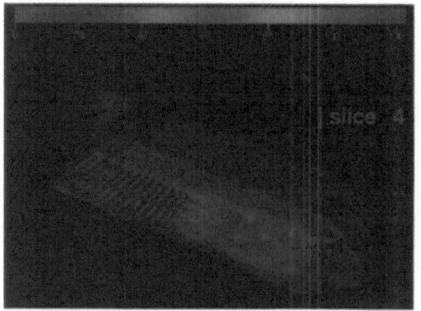

The vector field of the curl of the velocity computed on the fly by the algebraic manipulator (Leone and Scateni, Fig. 4)

Volume visualization of two chemical species concentration in a pressure field (Leone and Scateni, Fig. 2)

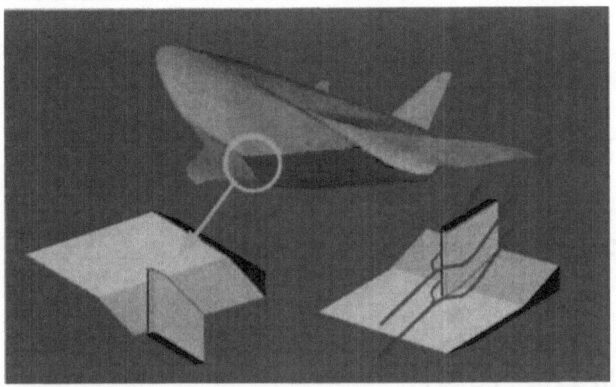

Schematic illustration of the geometry near a corner of an air-intake of a hypersonic transport aircraft (de Leeuw et al. Fig. 1) .

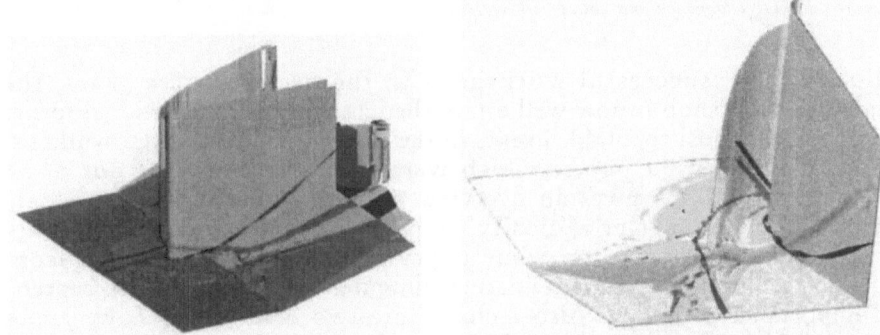

Two views of the flow field. Vortices are marked by a centered streamribbon and the position of shock waves is visualized by a transparent surface (de Leeuw et al., Fig. 2)

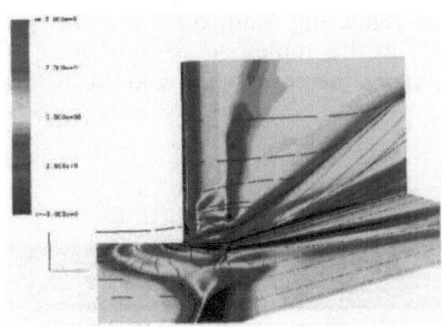

Colour coded distribution of convergence of the shear-vector field (de Leeuw et al., Fig. 6)

Colour coded convergence distribution blended with oil-flow pattern (de Leeuw et al., Fig. 7)

Patrick M. Hanrahan, Werner Purgathofer (eds.)

Rendering Techniques '95

**Proceedings of the Eurographics Workshop
in Dublin, Ireland, June 12–14, 1995**

1995. 198 figures. XI, 372 pages.
Soft cover DM 118,–, öS 826,–
ISBN 3-211-82733-1

(Eurographics)

Prices are subject to change without notice

Following five successful workshops in the previous five years, the Rendering Workshop is now well established as a major international forum and one of the most reputable events in the field of realistic image synthesis. Including the best 31 papers which were carefully evaluated out of 68 submissions the book gives an overview on hierarchical radiosity, Monte Carlo radiosity, wavelet radiosity, nondiffuse radiosity, and radiosity performance improvements. Some papers deal with ray tracing, reconstruction techniques, volume rendering, illumination, user interface aspects, and importance sampling. Also included are two invited papers by James Arvo and Alain Fournier. As is the style of the Rendering Workshop, the contributions are mainly of algorithmic nature, often demonstrated by prototype implementations. From these implementations result numerous color images which are included as appendix.

The Rendering Workshop proceedings are certainly an obligatory piece of literature for all scientists working in the rendering field, but they are also very valuable for the practitioner involved in the implementation of state of the art rendering system certainly influencing the scientific progress in this field.

Springer-Verlag Wien New York

Sachsenplatz 4–6, P.O.Box 89, A-1201 Wien · 175 Fifth Avenue, New York, NY 10010, USA
Heidelberger Platz 3, D-14197 Berlin · 3-13, Hongo 3-chome, Bunkyo-ku, Tokyo 113, Japan

Martin Göbel, Heinrich Müller, Bodo Urban (eds.)

Visualization in Scientific Computing

1995. 150 figures. VIII, 238 pages. ISBN 3-211-82633-5
Soft cover DM 118,–, öS 826,–. (Eurographics)

Visualization is the most important approach to understand the huge amount of data produced in today's computational and experimental sciences. Selected contributions treat topics of particular interest in current research, for example visualization of multidimensional data and flows, time control, interaction, and volume visualization. Readers may profit in getting insight in state-of-the-art techniques which might help to solve their visualization problems.

Wolfgang Herzner, Frank Kappe (eds.)

Multimedia/Hypermedia in Open Distributed Environments

Proceedings of the Eurographics Symposium
in Graz, Austria, June 6–9, 1994

1994. 105 figures. VIII, 330 pages. ISBN 3-211-82587-8
Soft cover DM 118,–, öS 826,–. (Eurographics)

This book represents the results from the Eurographics symposium on "Multimedia/Hypermedia in Open Distributed Environments", June 6–9, 1994, Graz, Austria. Its six sessions "Standards and Standards Exploitation", "Demonstrations", "Tools", "Hypermedia and Authoring", "Architectures", and "CSCW and Information Services" give a comprehensive overview about current research and development, including the future mm/hm standards MHEG and PREMO. The reader will profit in getting up-to-date information about the current trends in (the development of) mm/hm services and applications in open, distributed environments.

Prices are subject to change without notice

Springer-Verlag Wien New York

Sachsenplatz 4–6, P.O.Box 89, A-1201 Wien · 175 Fifth Avenue, New York, NY 10010, USA
Heidelberger Platz 3, D-14197 Berlin · 3-13, Hongo 3-chome, Bunkyo-ku, Tokyo 113, Japan

Springer-Verlag
and the Environment

WE AT SPRINGER-VERLAG FIRMLY BELIEVE THAT AN international science publisher has a special obligation to the environment, and our corporate policies consistently reflect this conviction.

WE ALSO EXPECT OUR BUSINESS PARTNERS – PRINTERS, paper mills, packaging manufacturers, etc. – to commit themselves to using environmentally friendly materials and production processes.

THE PAPER IN THIS BOOK IS MADE FROM NO-CHLORINE pulp and is acid free, in conformance with international standards for paper permanency.